EVOLUTION

IS

WRONG

EVOLUTION

IS

WRONG

*A Radical Approach to the Origin
and Transformation of Life*

ERICH VON DÄNIKEN

This edition first published in 2022 by New Page Books, an imprint of Red Wheel/Weiser, LLC
With offices at:
65 Parker Street, Suite 7
Newburyport, MA 01950
www.redwheelweiser.com

ISBN: 978-1-63748-005-2
Library of Congress Cataloging-in-Publication Data available upon request.

Cover design by Sky Peck Design
Cover art Ernst Haeckel, Kunstformen der Natur, 1904, "Discomedusae"
Interior by Maureen Forys, Happenstance Type-O-Rama
Typeset in Aviano, ITC New Baskerville and Weiss

Printed in the United States of America
IBI

10 9 8 7 6 5 4 3 2 1

TABLE OF CONTENTS

LETTER TO MY READERS

Dear Readers!

Evolution is a fact and there isn't the slightest doubt about it. This is basically what the world-famous *Encyclopedia Britannica* says, and this is exactly how the overwhelming majority of experts in the scientific community see it as well. Is the majority right?

Of course there is evolution with a small e. Everyone knows several types of dogs and knows that they all descend from a wolf-like primordial dog. But there are no pig dogs—that is, dogs with a pig's head—just as there are no mixed breeds of giraffes and lions.

But there are beings that live on our planet that, according to the evolutionary principle, should not exist. Can you, dear reader, imagine a carnivorous plant whose trapping leaves collapse in a fraction of a millisecond? But that do this only after the victim has touched two different bristles? Or, even better, a frog species that carries its young to term in its stomach? In the same stomach that was actually created to digest its food?

This is the kind of thing this book is about. This, and the controversial views of some scientists in their specific fields.

Cordially!

Erich von Däniken

EVOLUTION
IS
WRONG

CHAPTER 1

THINGS ANIMALS
ARE CAPABLE OF

PARASITIC SPECIES

Imagine the following scenario:

In flight, a wasp heads for its victim—a spider. The wasp stabs the spider in the back from behind and injects poison into the wound, paralyzing it. The wasp then lays an egg in the damaged part of the spider's body. The larva develops inside, and once it hatches, the newborn feeds on the host's innards. It begins by gradually devouring those parts of its victim that are not important for the spider's survival. This way, the spider stays alive and fresh as long as possible.

This scenario plays out every day. In Australia everyone knows of the so-called Captain Cook or redback spider-hunting wasp (*Agenioideus nigricornis*; see Figure 1.1). This small, annoying insect reproduces with the help of the poisonous redback spider. As with the well-known black widow, the redback spider's bite can be fatal. Because of this, every year in Australia, hundreds of people are treated with a serum after they are bitten by this spider.

Figure 1.1: The redback spider-hunting wasp (*Agenioideus nigricornis*)

Of course, we know that venomous spiders kill other lifeforms, but who kills the venomous spiders? Entomologists Patrick Honan from the Museum of Victoria, Australia, and Lars Krogmann from the State Museum of Natural History in Stuttgart, Germany, asked this question. The results were astonishing. Both the little redback spider-hunting wasp and its larger conspecific spider wasps (*Pompilinae*) attack venomous spiders and use the spider's bodies as breeding grounds for their young.

Several other species of wasps have mastered the art of abusing foreign hosts. The braconids (*Dinocampus coccinellae*) manage to manipulate ladybugs. Similar to the pompilids, the braconid wasp lays an egg in its victim's abdomen, and the wasp larva feeds on the beetle's body fluids. As soon as the larva has reached a certain size, however, it crawls out of the ladybug and pupates on the beetle's abdomen. This makes the ladybug mostly motionless, but its legs still twitch. After the new braconid wasp hatches, the ladybug recovers and can even reproduce again. Obviously, the wasp realm has developed phenomenal methods of feeding its brood.

The jewel wasp (*Ampulex compressa*; see Figure 1.2) ambushes cockroaches (*Periplaneta americana*; see Figure 1.3). Even though these roaches are ten times larger than the wasp, this doesn't prevent the wasp from suddenly jumping out of hiding and paralyzing its prey with the first sting. It aims the second sting directly at the roach's nervous system, precisely at the region that controls its escape capability. The wasp then uses the cockroach's antenna to direct its prey to a hole in the ground and, like a zombie, the cockroach moves on its six legs to its own grave. Once the roach is there, the wasp lays an egg in the cockroach and then builds walls around it with small stones. After four weeks, a new jewel wasp hatches out of that prison and looks for its next victim. The cockroach is dead.

Figure 1.2: Jewel wasp (*Ampulex compressa*)

Figure 1.3: American cockroach (*Periplaneta americana*)

The ichneumonids (*Ichneumonidae*) are masters of an even more perfidious procedure. The ichneumonid's host is the well-known orb web spider (*Plesiometa argyra*). The process is similar to the previous stories. First the parasitic wasp paralyzes this spider with a sting. The wasp then lays an egg in the spider's abdomen. The larva feeds on the spider's innards and grows. As soon as the larva is ready to hatch, the wasp, which has to stay close at all times, injects a new poison into the spider, which changes its behavior. Instead of weaving an orb web, as is innate, the spider begins to use its threads to wrap a cocoon around the larva. When this cocoon is ready, the parasitic wasp kills its victim and eats it. The wasp larva continues to grow and finally leaves its cocoon.

Speaking of orb weaver spiders, Darwin's bark spider (*Caerostris darwini*; see Figure 1.4 and Image 1 in the color insert) produces spider webs with threads of up to 25 meters (82 feet) in length. How? The spider positions itself in an air draft and lets the wind carry its silk across a stream or pond. The spider then climbs over its construct and affixes a new anchor thread to a place next to it. It then lets itself fall—spider bungee jumping—and be carried by the wind to the other bank. It repeats this game, starting from two main threads, until it has created a gigantic net that lies over the body of water. The most amazing thing about this construction is not the size of the web, but the strength of the spider's threads. The silk is ten times stronger than Kevlar (synthetic fibers used not only in jeans made for motorcyclists but also in bulletproof vests). In fact, the threads of Darwin's bark spiders are the toughest biomaterial in the world. Not only are they incredibly strong, they are also extremely light and thin. The mysterious evolution or desire of the spider must have prompted this: if they are going to make such long threads, they must make them stronger than the silks of all other conspecifics.

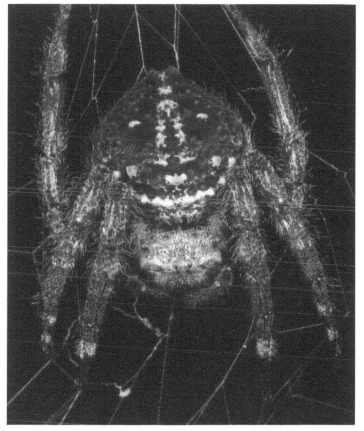

Figure 1.4: Darwin's bark spider (*Caerostris darwini*)

These examples raise perplexing questions: What was going on when the first jewel wasp pounced on a cockroach ten times its size? How did it know that its sting had to hit the exact point in the nervous system that controlled its victim's ability to escape? The wasp could not have known the structure of the cockroach's nervous system.

If the sting missed, the cockroach would have killed the much smaller wasp. And how was its experience passed on to the next generation? And which chemical brew was necessary to change the brain of an orb weaver spider in such a way that, instead of a web, it wove a protective cone around an alien being? What is the evolutionary mystery that creates the right mix of chemicals in the wasp's body?

And what about the host animals? Is what concerns them all normal and nothing extraordinary? "Nature" (I will come back to this later), in fact, knows innumerable parasites that need host bodies to feed their brood. After all, all so-called neuroparasites abuse their hosts' nervous systems for their own purposes. The flatworm or fluke (in the class *Trematoda*), for example, is a neuroparasitic monster with male and female reproductive organs. In other words, the worms can fertilize each other as well as themselves. And for them, one host is not enough to go through their lifecycle. After the worm has moved on to suckle on another animal, the first host lays eggs at some point. These get into the water, and larvae, the *miracidia*, hatch from them. They swim around until they are either eaten or encounter a special snail. In the latter case, the miracidium bores into the skin of the snail and grows into a brood tube. After several metamorphoses, daughter sporocysts develop from these tubes and these afflict the midgut gland of the snail. This is where rod larvae develop, which in turn produce tail larvae. From this, cercariae grow, which leave the host snail and enter a new intermediate host.

Other flukes use caterpillars as hosts. These caterpillars are subsequently eaten by birds. The eggs of the fluke are spread through the bird droppings and the cycle starts all over again. This all seems very simple, right? But how does the fluke know that the caterpillar it is abusing as a host is being eaten by a bird and that the bird droppings guarantee the survival of its species?

We have all heard about the tapeworm (*Schistocephalus solidus*), but who is aware that the larva of this disgusting animal needs a bird to mate? Even the beginning of the tapeworm's lifecycle is spooky. A tiny crustacean (a copepod) eats the larvae of the later tapeworm. This copepod, in turn, has to be eaten by a small, three-spined fish of the stickleback type. The stickleback then literally offers itself up to be eaten by a bird. These larva can only grow in this three-spined stickleback—the whole process does not work with any other fish. And at some point the larva have to get into a bird to mate.

The birth of the parasitic crab hacker barnacle (*Sacculina carcini*) seems just as impossible. This barnacle abuses a crab to help its offspring. How does it work? After a process of insinuating itself in its host, it forms a small sack on the crab's abdomen in which its own kind of eggs can grow. The male crab hacker barnacle then fertilizes these eggs, and the crab tends and guards the foreign brood in this sack as if they were its own. In cases where this barnacle infects a male crab rather than a female, another miracle happens. The parasite affects the hormones of the originally male crab and causes it to mutate into a female organism! Abracadabra.

Even ants can be controlled by parasites. It seems that for nature—whatever *nature* is supposed to mean—even the impossible is possible. A parasite called the lancet liver fluke (*Dicrocoelium dendriticum*) attacks grazing animals such as cattle or sheep. Its larvae are excreted in the feces of these animals. Snails then feed on this excrement and develop cercariae, which enter the snails' respiratory system. The snail spits out the cercariae in tiny slime balls. This slime is then eaten by ants, and it is only at this point that the original lancet liver fluke becomes active. This mucus enters the ant's nervous system and changes it completely. The ant positions itself on the tip of a blade of grass and waits to be eaten by a grazing animal. If this does not happen during the first day, the ant returns to its burrow and repeats its behavior until it is swallowed.

Well, you might be thinking, that's just what animals do. But you may be surprised to discover that parasites even control humans.

In the journal *Spektrum der Wissenschaft* (Spectrum of Science), biologist Sabrina Schröder drew attention to a parasite that can change humans.[1] The animal is called *Toxoplasma gondii* and was discovered in Tunisia as early as 1907. The parasite triggers the disease toxoplasmosis, which in turn changes the behavior of humans and animals. Toxoplasmosis is now known worldwide and is mainly spread through cat feces. If a rodent—for example, a mouse—ingests *Toxoplasma gondii*, it loses its innate fear of cats and literally offers itself to its archenemy to be eaten. Neurologists suspect that this parasite can cause diseases such as schizophrenia in humans. Studies have shown that people with toxoplasmosis are more prone to depression and suicide. In addition, toxoplasmosis can lead to inflammation of the brain (encephalitis). Humans are infected through exposure to cat excrement.

What evolutionary processes must these (and many other) animals have gone through? Think of the first wasp that attempted to fly onto a highly venomous spider. Spiders are clever opponents who defend themselves not only with their claws and by sucking the life out of their prey, but also with their sticky threads. Why did a wasp come up with the life-threatening idea of attacking a venomous spider—millions of years ago, if you like? At the time, there were, after all, enough other, more harmless lifeforms crawling around on the ground. How did the wasp get the idea to lay its eggs in the wound of a completely alien species? After all, the spider was not one of its related species that could be entrusted with its brood. And how does the wasp larva in the bowels of the spider "know" which innards it has to tap in sequence so that its host remains fresh and alive for as long as possible? Where does the wasp's offspring get its information from? And basically, in what way is this cycle supposed to have started? How did the first wasp egg get into the body of the venomous spider? Which came first? The chicken or the egg? The first wasp or the first wasp larvae in the spider's belly? And why is the process so cumbersome anyway? Wasps could lay their eggs anywhere; why, of all places, in the body of a living, venomous spider?

DANGEROUS SPIDERS

Over hundreds of millions of years, as the theory of evolution teaches us, around forty thousand different species of spiders developed. All of them had to come from some primordial spider and all have since developed completely different capabilities. The Australian funnel-web spider (*Agelenidae*) is considered to be the most venomous spider in the world because it kills so quickly. It developed a venom that is only lethal to primates and insects, but not to animals such as rabbits or chickens. How did this strange venomous cocktail, that kills some animals and not others, come about?

Most of the spiders on our globe kill their prey with venom. First they catch their victim—often, but not always in the web—then they kill it. One of the large European spiders is the mighty green-fanged tube web spider (*Segestria florentina*; see Image 2 in the color insert). It lives mainly in narrow cracks, crevices in the rock, or on tree bark and grows up to 4 centimeters (1.5 inches) in size. Its venom is painful to humans, but not fatal. The same is true for the Goliath birdeater (*Theraphosa blondi*). It can grow to be 12 centimeters (over 4.5 inches) tall and weigh 200 grams (more than 7 ounces). It can evoke fear for sure—but is not dangerous to humans. Nor does it hunt birds, as its name might suggest; rather it preys on insects and vertebrates such as mice and frogs. The same applies to the European tarantula wolf spider (*Lycosa tarantula*). It looks just as terrifying as the Goliath birdeater spider and can inflict painful bites on people, but they are not fatal.

Spiders . . . spiders . . . spiders with different hunting behavior and various weapons. And they are all related to one another. A water spider, also called a diving bell spider (*Argyroneta aquatica*), breathes air but then dives under water and carries the air with it in a breathing bubble. In terms of evolution, this is about as perverse as the whale, a mammal that lives in water. Spiders feed on land. They are armed with all kinds of weapons to catch their food, often webs or venom. So what makes a spider create an air bubble and hunt under water? After all, there it can be easily eaten by fish.

A dangerous arachnid native to Germany is the yellow sac spider (*Cheiracanthium*). This microbeast does not build a web but instead hunts its prey at night. Its bite causes chills, vomiting, and circulatory failure in humans.

Why am I listing all these spiders? Well, I'm interested in their different behavior and their various weapon systems, all of which have arisen in the course of evolution—says the theory. After all, Charles Darwin (1809–1882) discovered the diversity of species more than 150 years ago and postulated the idea of "geographical variations."[2] Everything has a common ancestor but develops differently in different regions.

There are different forms of the black widow (*Latrodectus tredecimguttatus*) in Europe as well as in North and South America (see Figure 1.5). Their bites can all be fatal (though rarely are) to humans. In addition, many female black widows eat their male counterparts after mating. As a thank you for fertilization? With their venom, they not only kill beetles, but lizards as well. Attention: black widows feel comfortable under planks around construction sites, also on the underside of toilet seats! But at least evolution has provided a warning signal: the Mediterranean black widow wears thirteen fiery red dots on their dark bodies. With this the animal signals, I am dangerous! Do not touch! At least that's what the theorists think. But why does the deadly Brazilian wandering spider (*Phoneutria nigriventer*) not also carry this warning sign?

Figure 1.5: Female Mediterranean black widow (*Latrodectus tredecimguttatus*)

The brown recluse spider (*Loxosceles reclusa*) is considered the most dangerous venomous spider in the United States. Curiously, this little animal only has six eyes—in contrast to common spiders that have eight. Was evolution satisfied with six eyes? Wouldn't eight have been more prudent? And then there is the Brazilian wandering spider (*Phoneutria nigriventer*). It is one of the deadliest animals in the world. Its venom is twenty times stronger than that of the black widow. The spider can be very aggressive and attack its prey with one jump. Their bite leads to muscle paralysis in humans and, as after a venomous snakebite, sometimes causes the heart and lung muscles to fail, resulting in a horrible death.

All of these strange creatures leave us with questions. It is not only their different mixtures of venom, their strange hunting methods, the way they abuse strange animals and use them as breeding grounds, but it is also their often unspeakable origins.

METAMORPHOSIZING INSECTS

Butterflies and moths are a prime example of this, including the Atlas moth (*Attacus atlas*) with a wingspan of up to 30 centimeters (more than 11 inches; see Image 3 in the color insert). Like many insects, the Atlas moth has compound eyes, which are made up of around eight thousand smaller eyes. It also has two larger individual eyes. Its antennae are slightly splayed. With these features, the male can smell a female from greater distances. These characteristics are urgently needed, because the Atlas moth is focused on creating offspring quickly—it only lives for a few days after it become a moth and doesn't eat anything during that time period.

The Atlas moth, as is the case with its conspecifics, butterflies, does not simply come about through a "birth" of whatever kind. Before its short winged adult life, it has to go through three stages of development: egg, caterpillar, pupa. It is the final transition between pupa and moth that makes the lifecycle of the moth/butterfly very confusing in evolutionary terms.

When caterpillars emerge from butterfly eggs, they do the actual eating for the later adult butterfly. Each caterpillar consists of fourteen evenly lined-up segments with a head at the front. After hatching, caterpillars first eat their own shell and then proceed to eat seeds, needles, and leaves from various plants. They are downright voracious.

Some species of caterpillars do it slightly differently, however: they live with ants in peaceful harmony. These caterpillars secrete a sugary liquid that attracts ants. The ants climb on the caterpillar—not to kill them or to scare them away with their formic acid, but to get at this sweet food. The ants then drag the caterpillar into their burrow. There it takes on the smell of the ants. It lets the ants feed it and, at the same time, it produces its sweet juice. Finally, after the ants maneuver their guest into a small niche, it pupates while the ants protect it from predatory conspecifics.

The pupa stage—this stage in all insects, not just butterflies—contains a "magical" genetic program. The cells of the pupae are totally transformed while a physical reorganization process takes place. Pupae have neither legs nor wings, let alone eyes or genital organs. In order for a caterpillar to become a butterfly, a completely new body has to emerge. In the case of the Atlas moth, the entire shape of the animal has to change, eight thousand smaller eyes (the compound eyes) grow together with the two separate eyes, and all previous structures of the caterpillar disappear. Finally, a new lifeform hatches: the butterfly.

In various ancient cultures, the butterfly was considered a symbol of rebirth, and artists of the Christian world recognized in it the power of resurrection. The soul leaving the dead body was shown as a butterfly. Obviously, wise men already practiced scientific thinking thousands of years ago. For a long time they had observed what was happening in the lifecycle of the butterfly: how an egg was formed into a caterpillar and then transformed into a butterfly through the process of pupation. But they could not know anything about genetics and therefore could not possibly ask the questions of our time. Each program for developing a lifeform originates in the cell, and this cell contains the DNA code, the double spiral of deoxyribonucleic acid. It is the DNA that carries the genetic information. Without DNA, nothing happens in the cell. (I will come back to this.)

But where should the information come from that creates a completely new body out of the pupa? This information must already have existed in the egg. What causes "evolution" to create a lifeform like a butterfly by means of such a complicated path, and why does the adult form only live just a few days without eating anything? Why certain caterpillars let themselves be fed by ants might be explainable. It just worked out that way. But the genetic process that turns an egg into a caterpillar, then from a caterpillar to a pupa from which a completely new creature emerges is difficult to reconcile or explain in terms of evolutionary theory. The information in the DNA strand in the cell also contains "stop and go" commands. When—in the course of time—will the next molting of the pupa be enabled? When does the basic wing form arise, when do the yellowish wing tips form, when do the slightly fanned antennae develop? When do the mouthparts grow? How about the legs and the eight thousand eyes? The sensory organs? (Butterflies not only have eyes but also have ears.)

The message of the genes must not only be "read" in the right sequence, but the correct processes must also take place at the right point in time, otherwise deformities will occur. For example, several species of the blister beetle (*Meloidae*) are unable to fly despite their wings because their wings are too short. Obviously, the stop signal for growth came too early. (Incidentally, blister beetles owe their name to an oily liquid that leaks from the pores of the animal's leg joints, which can cause blisters when it comes in contact with exposed skin. This liquid acts as a defense against ants and other animals.)

As with butterflies, blister beetles go through several metamorphoses before they become adults. Their larvae develop in the nests of certain types of bees such as sand or fur bees. A pupa emerges from the larva and next to it— inexplicably—an empty dummy pupa emerges. Finally, a new blister beetle creeps out of the skin of the real pupa. And all of this has—quite clearly—been arranged by "evolution."

FRESHWATER TO SALTWATER AND BACK

Such wonderful evolutionary quirks also resulted in different subspecies of the salmon. There are nine such subspecies worldwide, but they all have one thing in common: they are born in fresh water, swim into the distant salty sea and, years later, return to their place of birth. There are salmon in both the Atlantic and Pacific. The female Atlantic salmon (*Salmo salar*) lay their eggs in the upper reaches of rivers. After spawning, the majority of the adult fish die. After they hatch, depending on the water temperature, the young fish stay in the river for one to three years during which time their eyes, gills, and dorsal fins develop. They become hunters. At this point a change in their bodies begin. The blue and red spots on their skin disappear and a shimmering color covers the animal. Their innards undergo a metamorphosis: the freshwater fish becomes a saltwater creature. Salmon researcher William Shearer found that most salmon "swim into the Atlantic towards West Greenland."[3] Why they do this is unknown.

The situation is similar with Canadian salmon. There are five different species, the largest of which is the Chinook salmon (*Oncorhynchus tshawytscha*). Adults swim around 1,300 kilometers, from the Pacific via the Fraser River and the McLennan River to Swift Creek near the city of Valemount (British Columbia, Canada), to the place where they were born.

Another salmon species—the sockeye salmon (*Oncorhynchus nerka*)—also has to swim up the Fraser River and overcome massive rapids at Hells Gate. As if under the influence of a drug or out of an inner rage, the fish jump, again and again, throwing themselves toward the waterfalls in order to overcome them with lightning-fast movements. At the top of the rapids, grizzlies and black bears are perfectly positioned. They grab the very fatty meal when the salmon jump in the air. What drives these fish? Although the entire American continent lies in between these species—from Alaska, Canada, and the US, to Central and South America, to Tierra del Fuego—both the Canadian and the Atlantic salmon have the same instinct. The fish in both bodies of water behave the same way. A salmon may have covered thousands of kilometers in its lifetime, but after years, it finds the mouth of its home river with impeccable certainty. Only in this river does it swim upstream with tremendous exertion and overcomes rapids to be devoured by a bear.

Which navigation system directs these fish? The Earth's magnetic field? In fact, cells containing magnetic iron oxide were found in the salmon mucous membrane. However, that does not solve the riddle. After all, the Atlantic (see Image 4 in the color insert) and Pacific salmon have lived separately from one another for millions of years. The magnetic fields in east and west are different. In addition, several magnetic pole shifts have been proven to have occurred during this timeframe. In other words, the magnetic field has reversed. And speaking of the magnetic field: it would never deviate in a narrow, limited geographical area to such an extent that zones with different strengths would emerge.

This means that several rivers that lie in one area and are sought out by salmon would hardly be distinguishable from one another for the animals. How do they know which is the right one for them?

Even the thought that the fish would orient themselves toward the stars isn't plausible. It is well known that the Earth is not a perfect sphere. It is slightly flattened at the poles—that is why it wobbles on its axis. In astronomy, this movement is called *precession*. In the course of 25,800 years, people and animals have seen different constellations again and again. Today the sun stands in front of the constellation of Pisces at the beginning of spring. In 6,450 years, it will be the constellation of Sagittarius, 12,900 years after that it will be the constellation of Virgo, then Gemini, and so on. It does not help to imagine that some form of the primeval salmon memorized a constellation and passed it down to all its subsequent generations. Salmon do not practice astronomy. And besides, what magic spell causes the animals to change not only their skin but also their innards so they can switch from fresh- to saltwater? I read somewhere that the animals swim down the rivers in search of food found in the very distant ocean. This is nonsense. The food in the river is just as abundant as that in the ocean. In addition, salmon—males and females—finally return to their (nutritionally weak?) home river.

It has also been suggested that it is the fish's sense of smell that directs them precisely to their home waters. I'm sorry, what? The rivers pour into the gigantic masses of water of the oceans. There, in the salt water, even the smallest special scent from a river 1,000 kilometers (620 miles) away is lost.

So the answers, so far, are unsatisfactory. Let's keep in mind that each female salmon releases around thirty thousand eggs into the water when she spawns. Every egg has to contain the genetic information, which in broad terms is to reach adulthood, to change appearance and organ structure, to swim to the distant saltwater ocean, to let the internal navigation computer develop, and—finally—to return to the place of birth. It should be noted that salmon did not acquire this information in an evolutionary process lasting hundreds of millions of years; no, the information was already in the genes of the first salmon. Otherwise, the lifecycle could never have started or continued.

AND THE OPPOSITE: THE EUROPEAN EEL

And why is the eel species programmed in the complete opposite way when compared to salmon? Let's look at it step by step. European eels (*Anguilla anguilla*) live in fresh water rivers, lakes, and lagoons for years and then swim to the saltwater Sargasso Sea, the place of their birth, some 6,500 kilometers (over 4,000 miles) away—the exact opposite of what salmon do. The Sargasso Sea, located between Florida, the Bahamas, and Bermuda, is larger than the Mediterranean. There, like salmon, female eels spawn and die. The males fertilize the spawn and tiny eggs grow. These become larvae that drift toward Europe with the ocean currents. Here, they mature into young eels, which fight their way up the rivers in large swarms. After a few years, they become sexually mature. Then their appearance changes.

The dark eel takes on the form of a shiny, silvery fish, adapted to the sea. The eels want to go back to their place of birth and swim down the rivers. Sometimes there are problems with this. In an article about eel migrations, Carsten Jasner writes: "If the path to the river is cut off because the animals live in ponds or pools, they quickly meander across land."[4] This is made possible by the animals' fantastic adaptation to their environment; they have special gills and can breathe through their (moist) skin. And, of course, they can exist in both fresh- and seawater.

As with salmon, attentive scientists are looking for the reasons for eel migration. Norwegian researchers working with marine biologist Caroline Durif placed the animals in closed tanks and exposed them to various magnetic fields. Depending on the magnetic field and water temperature, the eels oriented themselves in different directions. But what inner compulsion forces them, year after year, to leave their nourishing river landscape, which offers optimal conditions, and hone in on the Sargasso Sea so far away? With the huge distances and different ocean currents, the "smell of home" explanation doesn't help. The proposed explanations, whether they have to do with the availability of food, water temperature, or the magnetic field, also sound hollow. If these were plausible, countless other lifeforms would be affected as well, but they have no intention of adapting their behavior to that of the eels.

USING ELECTRICITY UNDERWATER: THE ELECTRIC EEL

Another snake or fish-like animal is the electric eel (*Electrophorus electricus*). This animal isn't actually an eel at all, even though the name suggests it (see Figure 1.6). Electric eels are actually marine fish. But—and this is where the evolutionary leaps begin—electric eels are air-breathers.

Figure 1.6: The electric eel (*Electrophorus electricus*)

The electric eel has to breathe air rather than breathe through water and surfaces every 10 minutes to do so, on average. They then exhale through their gills. Current marine biology differentiates between three different types of electric eels, which separated as early as the Miocene, roughly 23 million years ago. As their name suggests, electric eels have a terrifying weapon with which they can kill their prey—electric shocks. The world-famous explorer Alexander von Humboldt (1769–1859) described how indigenous people in the Amazon caught electric eels in his presence using a special method to avoid being shocked. But where does the electricity in the electric eel's body come from?

Biologist and neurologist Kenneth Catania from Vanderbilt University in Nashville, Tennessee, explored this question. He found that three chambers develop in the body of the electric eel that produce electricity with voltages of up to 860 volts. (For comparison: In our society, 500 volts is considered high voltage; you would only get near something with this voltage if you were wearing thick rubber gloves.) Each of these chambers consists of electricity-generating elements that produce electrocytes. Usually—according to Catania—only one chamber releases its electrocytes to the outside world. On the other hand, in the event of very strong electric shocks, all three chambers interconnect and produce up to six thousand electrocytes. This then produces the high voltage.[5]

Alexander von Humboldt personally experienced how the indigenous people in the Amazon rainforest drove horses into the water and how these animals were literally bombarded with electric shocks by the electric eels. Humboldt described how the horses literally sank under water, stunned by the strong, incessant electrical blows. He went on to claim, "Other horses, snorting with bristling mane and showing wild fear in their arrested stare, stand up again and try to escape . . ."[6]

Obviously, the strange evolution of these fish developed those fatal electric shocks before humans invented the electric chair. But how exactly? All three types of electric eels can use this electric weapon, some with a little more voltage than the others. But the command for the creation of these electricity-producing organs must have existed in the genome of the animals over 23 million years ago. Otherwise, not all three electric eel varieties, which are distributed across the Earth, would have the ability to generate high voltage. How these chambers are supposed to have developed over a process spanning millions of years remains inexplicable. Did it start with a tickle in their stomach? And if the animal touched another animal, why did the electric shock only affect their prey and not other electric eels as well? After all, both animals are in the water, and it's common knowledge that water is a conductor of electricity (even though pure water would not be) because most of the time it contains charged ions and impurities. In the course of evolution, did an electric shock suddenly go through an electric eel? If so, why didn't it die from it and break the chain of evolution? Shouldn't the first high-voltage shock have been a moment of tremendous horror for the electric eel itself? And how does the small brain of the eel determine whether it needs a very strong or weak charge of electricity to kill an enemy? And did it need electricity from just one or all three chambers? It would have to make such decisions at lightning speeds. If it doesn't decide quickly it might be eaten.

It is very hard for me to imagine what seems like a slowly developing power plant in the body of an electric eel. It cannot be that all electric eels developed their weapons independently of one another. After all, the entire species has mastered the feat. So, once again, it has to do with the genome, the millions-of-years-old message contained in the genes. The belief that evolution had an endless amount of time at its disposal for each variant is exactly what the word expresses: belief.

ON THE MOVE: SAFETY IN NUMBERS

Ever since we developed more and more sophisticated cameras, and brilliant, absolutely clever, and infinitely patient people began to use them, animal films have been produced that continually amaze us. One of these sensational films, produced by British documentary filmmaker Nick Stringer and his cameraman Rory McGuinness, is titled Tortuga and can also be admired in a cinema version.[7] It shows the life of loggerhead sea turtles (*Caretta caretta*), who return year after year to the beach where they were born to lay their eggs in the sand. Around two million of these turtles hatch on the beaches of Florida alone. It can take three days for the baby turtle to break free from the egg. The distance from the birth hole to surf is a death-defying path. The shell of the turtle has not yet hardened, and so the young turtles, by the thousands, are devoured by the lurking seagulls, other birds, coyotes, and large crabs. The turtles have no weapons. They rush hectically across the beach, over stones and branches. They seem to be driven by an inner panic. Don't ever rest—just take the straight path toward the surf. There they dive in—and are swallowed by the next predators, sea creatures of all kinds. The surviving turtles swim about 70 kilometers (over 40 miles) and rest on algae carpets.

Once rested, they swim thousands of kilometers through the North Atlantic to Newfoundland, and from there to the warmer waters of the Azores, and finally to the Caribbean Sea. They travel relentlessly for twenty years and eventually return to the beach where they were born. Other sea turtles (*Cheloniidae*) such as the hawksbill sea turtle (*Eretmochelys imbricata*) hatch on the Atlantic island of Ascension and spend most of their time at the Brazilian shore—until they too return to their place of birth—2,000 kilometers (over 1,200 miles) away (see Figure 1.7).

Figure 1.7: A sea turtle (*Cheloniidae*)

In addition to the hawksbill turtle, and the related loggerhead turtle—of which millions regularly hatch on Florida's beaches—there are seven different species of turtles on our planet (see Image 5 in the color insert). All are lung breathers. Their metabolism actually changes when they dive. While under water, their blood gradually takes on carbon dioxide (CO_2) and they have to surface every 30 minutes on average to replace the CO_2 with fresh oxygen. All turtle species originally descended from tortoises that waddled into the water at some point in the later Paleozoic Era—which spanned the period from 541 to 252 million years before our time. Paddles developed from their feet, and gradually—as always, over millions of years—their bodies are said to have adapted. Their shell changed, and the sea turtles lost the ability to hide their large heads under their shell in case of danger, which their relatives, the tortoises living on land, still do today. In addition to these other adaptations, to regulate the salinity, salt glands grew in the body of the sea turtle.

What happens when these animals' births, on Florida's coasts, repeated in other parts of the world, such as Borneo, the Philippines, Malaysia, or the Atlantic island of Ascension? As if by a supernatural command, tons of turtles suddenly emerge from the water. The females shuffle to the beach at night and dig a hole in the ground. They lay their eggs in the pit they've dug and covered them with sand. Over the next 60 days, the sun heats the holes and incubates the eggs. In the pit, at a temperature of over 29.9 degrees Celsius (86 degrees Fahrenheit) female turtles develop, and, at a smaller number, males as well.

It is well known that different animal species move in schools. And there are natural explanations for this group behavior. For example, gigantic swarms of sardines appear year after year along South Africa's coasts. Millions upon millions of fish bodies form into something that resembles a carpet, 1 kilometer wide and 12 kilometers long (between .6 miles wide and 7.5 miles long). The reason for this spectacle lies in the fact that two ocean currents converge at the Agulhas Bank, a shallow coastal strip on the edge of the southern African continental shelf. It is here that the warm water of the Indian Ocean mixes with the colder waters of the Atlantic. The sardine spawn have no choice but to be dragged along by the currents. A real mystery of the schools of sardines, however, remains their behavior toward one another. Although no lead animal transmits any commands, millions of fish move in unison or perform the same maneuvers at lightning speed without ever getting in each other's way. No sardine touches another. Have the little animals mastered something like telepathy? The formation of the school is not a mystery in this case—but the common behavior of the animals definitely is.

SWITCHING SEX: HERMAPHRODITIC CREATURES

The evolutionary explanations become tedious for lifeforms that can change their sex or that are hermaphroditic. I have already reported on the parasitic flatworms, known as flukes, which possess male and female sexual organs and can also self-fertilize and the male crab that mutates into a female. Then there are all the land snails—they have both female and male sex organs.

And then there are the clownfish (*Amphiprioninae*), one of which features prominently in the film *Finding Nemo* (see Image 6 in the color insert). Clownfish live in groups that are always dominated by a female. If the female dies, the strongest male changes into a female.

At times, omniscient evolution produced impossible forms. For example, barnacles (*Balanidae*) that cling to mussels, snail shells, or humpback whales. Once they have attached, the barnacles can no longer change their location, which remains linked to its host. But even barnacles are hermaphrodites—they can fertilize each other. Since the barnacle sticks firmly to its host, it has developed a giant penis that is eight times longer than itself. With this magnificent organ, it scans its surroundings for sexual partners, and eventually fertilizes the eggs. The barnacle eggs grow into larvae, from which a thin shell develops, which attaches to a host. Isn't there an easier way?

In Australia, biologists discovered a tiny sea slug (*Siphopteron quadrispinosum*) that, in relation to its body, also has a giant penis, which, in addition, splits into two tubes. One tube is pointed like a hypodermic needle. With this, the snail injects a paralyzing liquid into the body of its partner, a kind of knockout droplet. Then the second tube injects sperm. Evolutionary biologist Nico Michiels of the Eberhard Karls University in Tübingen, Germany, said: "That reminds strongly of a rape."[8]

Hermaphrodites like these often occur in nature. Their offspring are copies (clones) of themselves, but this does not prevent them from changing individually over the course of their lives. Hermaphroditic animals can multiply—that doesn't work for humans.

In addition to sexual reproduction, "nature" (I will come back to this later) knows asexual reproduction, popularly known as *virgin generation*. This occurs with wasps and bees as well as with aphids (sap-sucking insects) and some types of fish. The theory of evolution explains that the hormones trick the egg cells into thinking that they are fertilized, and as a result, they begin to divide. But how is that supposed to work? Do the hormones have something like spiritual vibrations that make wishes a reality?

MAMMALS OF THE SEA

Evolution is adaptation, change, and an endless amount of time that makes anything possible. Even the impossible.

There are ninety different species of whale on our planet, all of which are originally descended from an ancient whale. Whales are mammals. The ancestor of the primeval whale lived on land, and evolution theorists are convinced that this ancestor is closely related to the hoofed animals (*Artiodactyla*). Then, in the Eocene, 50 million years ago, the transformation from a hoofed animal to an aquatic animal began. You may wonder what links these two animals. It turns out that both animal species have a very similar ankle bone. The connecting link is a being called the walking whale (*Ambulocetus*). The corresponding fossils were discovered in Pakistan and have a length of 3 meters (9+ feet).

This walking whale could both swim and move on land, much like a seal, but climate change forced the animal to stay in the water more and more. Over the course of millions of years, hippos developed from this, and ultimately, from them, the whales. In short, this corresponds to the history of the development of the whales. It is also not without controversial opinions held among experts because bones of whales can be proven over a period of 50 million years—those of hippos only for 15 million years.

Whether it's the hippopotamus or some other ancestor, some animal has to be the forefather of the whales. Depending on the geographical latitude and water temperature, this forefather mutated into baleen whales (*Mysticeti*) or toothed whales (*Odontoceti*). The former filtered only plankton through their mouths, and yet they grew into giants of the seas, while the latter became predators and carnivores. The blue whale (*Balaenoptera musculus*; see Figure 1.8 and Image 7 in the color insert) is up to 33 meters (108 feet) long, weighs 200 tons, and is the largest animal on Earth. Around ninety offshoots with different properties have emerged from these two types, but they all have common properties.

Figure 1.8: Model of blue whale (*Balaenoptera musculus*)

Whales breathe air, as do their relatives, the dolphins (*Tursiops truncatus*; see Figure 1.9). Depending on the type of whale, they can stay under water from a few minutes up to two hours. The whales' infants are born with fully developed bodies. Since the birth takes place under water, the mother whale has to maneuver her young to the surface immediately, otherwise it will suffocate. After it has gotten its first breath of air, the baby whale looks for its mother's breast—under water, of course. Since the newborn whale has no lips with which to suckle, the mother injects her milk directly into her baby's mouth with strong muscle pressure.

Over the course of millions of years, the legs of a former hoofed animal (known as an ungulate) developed into fin-like appendages called flippers. In the middle of

Figure 1:9: Bottlenose dolphin (*Tursiops truncatus*)

the back, on which the former ungulate had nothing, an additional fin grew. Both the flippers and the fin are necessary so the animal can maneuver in the water. Without them, the whale would not be able to propel itself effectively or efficiently through the water. In the back of the huge body, a huge caudal fin formed from movable cartilage. The genitals of the male and the mammary glands of the female were drawn into the body in a watertight manner. The nostrils of the whales, which, in the case of the hoofed land animals (like the hippopotamus) were located in the front of the face, developed on the upper part of the head and mutated into the blowhole. Elongated, streamlined heads also developed. A thick layer of fat, called blubber, grew under the animals' skin. A fine friction surface developed on the outer skin that prevents the water from vortexing. In fact, tests in large water tanks and aquariums have shown that dolphins have a more effective friction surface than objects made by humans with the same streamlined shape!

Evolution gave rise to lungs in whales that can process 90 percent of the amount of oxygen absorbed. With humans it is only 15 percent. Depending on the type of whale, the animals can reach diving depths of 100 to 3,000 meters (328–9,842 feet). That's 3 kilometers (almost 2 miles) under water! Sperm whales reach these depths without any problems and can also stay under water for up to 90 minutes. Their relatives, the beaked whales, can remain submerged for up to 2 hours! Evolution created a system of squeaking tones for whales to use to communicate with one another; people call them whale songs. These tones are supposed to be audible in the water for several hundred kilometers. Researchers have analyzed a total of 622 different tone sequences at different frequencies. And one more strange thing: sperm whales can live to be up to 100 years old, and bowhead whales even reach an age of 200 years.

Actually, all of these mutations should be enough to pester the brilliant theorists of evolution with questions. But it gets even better. Among the whales there exists an extraordinary specimen, the narwhal (*Monodon monoceros*). This species is only around 5 meters long (~16 feet) and weighs almost 2 tons. Narwhals prefer the cold. Their hunting area is in the Greenland pack ice. The narwhals have something that their relatives lack: a tusk that is up to 3 meters (almost 10 feet) long and weighs 10 kilograms (22 pounds). This developed from the left canine tooth of the upper jaw of the male and is twisted in a helical counterclockwise direction. Narwhals use their tusks as weapons, sometimes against each other. For centuries, this ivory tusk was collected as an exquisite trophy and accordingly compared in value to gold. Leaders of churches and royal houses, but also museums, bought these narwhal teeth.

The narwhal's tusk was considered to be the ivory of the sea unicorn and was said to have various healing powers. In the Venus Church in Venice you can still marvel at two narwhal teeth that were stolen by the crusaders in Constantinople. And in 1671, the Danish King Christian V (1646–1699) was crowned on a throne of narwhal teeth. A 2.70-meter-long (almost 9-foot-long) example of a narwhal tooth is exhibited at the German Leather Museum in Offenbach, Germany (on the Main river).

FURTHER QUESTIONS

Why, of the ninety species of whale, did only the narwhal develop a tusk? Evolution makes interesting leaps. And it raises legitimate questions:

Cells in the skin of the salmon contain components of magnetic iron oxide. This is the reason that salmon can orientate themselves on the magnetic field. But the Earth's magnetic field is constantly changing. Which existed first? The mucous membrane with iron oxide or salmon migration? And what miracle steers the eels in the exact opposite direction?

What forces loggerhead sea turtles to return to their birth place on the beaches of Florida, even though tens of thousands of them are killed and eaten there while the related green sea turtles aim for a completely different destination: the Atlantic island of Ascension?

The copepod, a small crustacean, has to eat the larva of a tapeworm. In turn, the copepod must be devoured by a fish called the three-spined stickleback. Only inside this fish does the larva of the tapeworm grow. This larva is then excreted and must enter a bird's intestines. How should the tapeworm, then the copepod, then the three-spined stickleback and then a bird know this chain of development? If one single link in this chain is missing, the tapeworm can't develop.

The same goes for the little liver fluke. How should it know that its larvae are only passed on through the droppings of grazing animals? And if it doesn't know—as evolution theorists assume—why does it do it?

The Darwin's bark spider weaves giant webs with filaments that are up to 25 meters (82 feet) long and stronger than Kevlar. Which mind is supposed to have programmed the chemistry in the spider glands in such a way that these threads can arise? Which came first: the length or strength of the thread? One doesn't work without the other. A thread with half the length did not reach across the pond, and the first long thread did not have the necessary pulling power.

Which evolutionary engine gave both sardines and herrings the ability to make exactly the same moves by the hundreds of thousands within a tenth of a second?

All turtles are lung breathers. Diving is against their nature. A slow adjustment of their lungs, which would enrich their blood with oxygen under water, is not possible.

Evolution should create advantages for the specific lifeform. But the sea turtles can no longer pull their heads under their shell when they are in danger—an evolutionary step backward compared to the land turtle.

The impact of the Chicxulub asteroid, according to science, killed all species of dinosaurs 66 million years ago. The crater is located on Mexico's Yucatán peninsula. Why didn't the heat or the gases from the asteroid kill all animal species? After all, they affected the entire globe.

Climate change drove the ungulates into the water. And what about the countless other land animals? Climate change didn't just affect the ungulates.

The Australian sea snail has two penises. Which one developed first? Could the snail shed its semen only after the first penis injected paralyzing drops?

Immediately after their underwater birth, whale calves must be maneuvered to the surface, otherwise they will drown. This behavior cannot have developed slowly—otherwise there would be no more whales. The offspring would have all drowned during birth. Whale mothers inject their milk directly into the mouths of their young. This happens under water and using the pressure of the muscles. How is this process supposed to have developed slowly?

Water spiders maneuver an air bubble under water in order to be able to breathe there. Did a primeval water spider ever exist that developed this behavior? How is it supposed to have passed this on to its descendants?

Contradictions? Inconsistencies? Following Darwin's basic idea, evolution theorists explain everything using time: the millions of years that development had time to adapt. Evolution played around and experimented to see if something was right. But half-lungs, quarter-wings, an unfinished penis, or just a twentieth part of a magnetic field that is perceived do not work. Neither does a school of herrings or sardines, in which only half of the animals perform their lightning-fast, synchronous movements. The command "to the right, a quarter to the left, 2 meters up" affects all animals in the entire school. And the millions of fish in other schools practice the same behavior. So it's innate. Accordingly, the message must have already existed in the fish spawn—genetic information, a million times over, through the ages.

PISTOL SHRIMP

There is a creature named the pistol, or snapping, shrimp (*Alpheidae*). To date, 36 different species are known, which are further subdivided into 600 subspecies. All animals are descended from a primal crab and live in the tropics, occasionally in brackish water, but also in coral reefs and the deep sea. The animals are more than just hermaphrodites because the males can not only transform into females, but from females they can turn back into males (see Figure 1.10). Biologist Emmett Duffy, from the Virginia Institute of Marine Science in Gloucester Point, Virginia, was even able to prove that this type of crab, similarly to ants, forms veritable communities, much like nation states.[9] What is extraordinary about these animals is their ability to very quickly produce a loud bang using a jet of water, which at the same time causes a flash of light. That bang is 200 decibels. For comparison: the noise of a jet is around 120 decibels. A bubble filled with steam—the so-called cavitation bubble—explodes at a temperature of up to 470 degrees Celsius (878 Fahrenheit). It is logical that every attacker immediately takes flight after this loud bang and light effect. What genetic changes were necessary to develop this weapon in stages?

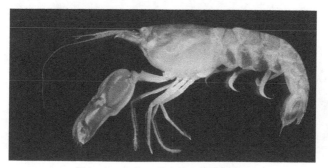

Figure 1.10: Pistol shrimp (*Alpheidae*)

CHEMICAL WEAPONS:
THE BOMBARDIER BEETLE

Evolutionary biology has plausible explanations for the origin and behavior of innumerable kinds of species, but definitely not for all. Here is an example that I have brought up in an earlier book.[10]

Among all the beetles on Earth, there is a phenomenal creature: the bombardier beetle (*Brachininae*) (see Figure 1.11). It scares and kills its enemies with a venomous mixture that shoots out of an explosion chamber in its abdomen at 100 degrees Celsius (212 degrees Fahrenheit). The secretion is composed of the chemicals hydroquinone and hydrogen peroxide as well as the enzymes catalase and peroxidase. This mixture produces a catalyst made from toxic benzoquinone. As soon as the bombardier beetle feels threatened, this mixture explodes between its legs and shoots specifically at the attacker's face. The different chemicals enter the explosion chamber through very thin tubes. After each explosion, this chamber closes again immediately, so it is ready to facilitate the next explosion. Even more amazing: two tiny

Figure 1.11: Bombardier beetle (*Brachininae*)

reflectors, one on each side, ensure that the beetle can even shoot around corners. Its venom not only kills small attackers, but also toads that are hundreds of times its size.

How did such a weapon develop slowly in the beetle's body? The wall of the explosion chamber consists of a type of chemical-resistant coating made of proteins and chitin. How should these chemical building materials have slowly formed in the body over many generations? Was there an empty explosion chamber first and did the ingredients then trickle in? When did the gland develop that can be steered into an exact target direction on the beetle's command? When did the side reflectors develop? And the explosion chamber with its two valves? In fact, it should have been there before the chemicals formed. After all, the ingredients must have been created in separate chambers, otherwise the beetle itself would have exploded when the chemicals came together—evolution would have been aborted. What about the tiny flap that opens quickly before each shot and then closes again immediately? If even 1 gram (.035 of an ounce) of the mixture were to trickle onto the beetle's skin, it would die.

And then think of the little animal's mind, coupled to its sensory organs—the eyes, its sense of taste, the sensors that perceive the vibration of its opponent—the mind has to give the order for the chemicals in the different chambers to exit and mix. When an enemy appears, this needs to happen in the blink of an eye. At the same time, the gland expelling the venom must be aimed at the enemy like a cannon barrel. Then the command to shoot must be given. By which "state of mind" did the individual parts of this weapon system come together without mutually eliminating one another? Did the beetle wish its body could now develop two glands with different chemicals in

addition to the existing internal organs? Did it also imagine adding an explosion chamber with a cannon hatch
that opens and closes in a flash? And two outer reflectors
made of very resistant material that not only have to withstand the aggressive acids, but also temperatures of up
to 100 degrees Celsius? Just a single chemical chamber
would have been useless. The beetle needs two to mix
the venom. And a gland, aimed at the target but *without*
the valve, which opens and closes in a flash, would not
have allowed the chemicals to leave the body. The entire
system only works in perfect cooperation. How could the
different components of this catapult develop slowly and
gradually?

GETTING AROUND AS A BAT

An even more extreme example are the phenomenal
sensory organs of bats (*Chiropter* of which the Pipistrelle
bat is a member; see Figure 1.12). There are more than
one thousand subspecies around the world that have
existed for about 50 million years. With the exception of
Antarctica, bats are widespread on all continents. Their
sexual behavior is strange. The animals live in groups and,
in order to sleep, they hang upside down from the ceil-
ing using the hook-like claws on their feet. For bats that
hibernate, mating takes place just beforehand. For some
bats, in order to mate, a willing male flies toward a female
in the dark and clasps her. A bite on the neck wakes the
female and the two mate. There is no parading around or
mating dance of any kind. After copulation, the female
begins hibernating, but—and this is crucial—fertilization
of the egg does not take place immediately. It takes place
weeks or months later when hibernation has ended. This

mechanism prevents the young from being born during the cold winter. Strange. An evolutionary process keeps the sperm in a waiting chamber until the time is right? Sometimes for days, weeks, or even months?

Figure 1.12: Pipistrelle bat (*Pipistrellus pipistrellus*)

The tracking system of bats is absolutely sensational. They emit several tones in short sequences, including those lasting one-hundredth of a second, at frequencies between 9 and 200 kHz. What is special about it? Other animals, from whales to dogs, also make sounds. But the sounds emitted by bats are reflected by insects, including tiny fruit flies just 3 millimeters in size. And bats have movable ears that they can turn and tilt. If the left ear registers an echo a fraction of a second earlier than the right one, the winged mouse knows in which direction to fly. The animal's whole body specializes in echo sounding. The bat manages to sort all the signals in the correct order and to give its wings the commands to change its flight path at lightning speed. Bats' brain also calculates the shifts in the sound due to the Doppler effect, which is created by moving objects. (Think of how the drawn-out horn signal of a car sounds different when the vehicle is approaching than when it is moving away.) Experiments with bats, whose eyes were taped shut, gave astonishing results: the bats flew around even the smallest obstacles without being able to see, even when the object was a stretched-out wire a quarter of a millimeter thick. The bat's brain processes the incoming echoes not in some unsystematic way, but in the correct *chronological* order of their occurrence. How far away is the target object? How big is it? In what direction is it moving? Toward me or away from me? Is there a wall in the background or is there open space? Say a bat is in an attic. Which obstacles—book shelves, curtains, pictures, wiring—change flight behavior? The analysis of the incoming data takes place within milliseconds. It should always be remembered that both the bat and its target meal are in flight. The incoming information changes every hundredth of a second.

Our radar systems register aircraft at a distance of 300 kilometers (186 miles). They capture every movement of the target object, even up to a distance of 1 kilometer (~ half a mile). But we do not have such sophisticated technology to capture and track a 1 millimeter target in a narrow space. An attic may have a dimension of 20 meters (65 feet) in length and width with a height of 3 meters (~10 feet) (i.e., a volume of 1,200 m^3[1215 sq. feet]). The bat conquers both space and target. Compare its abilities with the calculation capacity of a computer that has to continuously calculate new target positions in tenths of a second because both prey and hunter move continuously in the previously mentioned space—not just in one direction, but forward, backward, up, down, sideways, circling around obstacles, and taking into account the size, speed, color, and temperature of the prey and the hunter. We don't possess any comparable technology. However, evolution makes it possible.

Evolution has also created a subspecies to the normal bats: the vampire bat (*Desmodontinae*). These are the only mammals that feed entirely on the blood of other mammals; they can also climb vertical walls. Like their relatives, these animals send out signals and measure the size and body temperature of their prey. A vampire bat flies silently to a cow, for example, and moves toward the area of skin under which a vein is located. It then licks the skin over the vein with saliva, which contains an anesthetic substance, so the victim doesn't feel anything. The vampire bat then removes the hair and bites off a piece of skin and sucks up the welling blood. The saliva of the vampire bat also contains an anticoagulant substance so the blood remains thin and running normally.

Evolution theorists are convinced that vampire bats originally tapped birds and switched to the blood of mammals over the course of millions of years. Perhaps. But evolution must also have developed the sickle-shaped canine teeth that are necessary for cutting through skin—molars for chewing do not exist—and, of course, two chemicals: one that numbs the area and one that prevents the blood from clotting. In addition, two sensors had to be "constructed" that pinpoint the exact spot under the skin where a vein is located. Mammals often have thick fur, after all.

BLENDING IN AND REGENERATION: THE CHAMELEON AND SALAMANDER

Imagine you just clearly saw a small animal and then it practically vanished into thin air in front of your eyes. The animal is called a chameleon (*Chamaeleonidae*), and the play of colors on its skin can make it almost invisible when it adapts so that it almost blends in with its surroundings. Patient and extremely clever researchers have studied the life of chameleons and have come to astonishing conclusions.[11] Chameleons can change their appearance and suddenly look like leaves or branches that even vibrate in the air and adapt their color (see Image 9 in the color insert). How does it work? Different layers of skin make it possible. The bottom layer contains the crystalline dye guanine. This is ionizing, can refract incident light, and cause iridescent effects—comparable to a soap bubble in sunlight. On top of this is a layer of skin with cells that contain black-brown melanins. Melanins are pigments that determine the color of the skin or feathers. The third, topmost layer of skin contains so-called xanthophores, which are red and yellow dyes. The chameleon manages to activate all three skin layers in a targeted manner. The result is not just any arbitrary pattern but exactly what the chameleon wants. If the animal wants to look green, it will bring out the yellow and blue pigments. To become dark, melanin is pushed to the surface. If the chameleon is exposed to high levels of solar radiation, it changes color by absorbing ultraviolet light. The small animal also adapts to changing temperatures: when a chameleon gets cool, it takes on a dark color, and at high temperatures, a light color. In situations of danger, the color change takes place within a few seconds. It's not just about becoming green, blue, or dark. The chameleon imitates the appearance of leaves or branches that it happens to be crawling along on at that moment. If the leaf moves in the wind, the skin adapts to the changing pattern. The animal is difficult to see for an attacker.

There are around 200 different species of chameleons, which, it is said, originally split off from a primal family of iguanian lizards, the *Agamidae*, in the Upper Cretaceous Period, 100 million years ago. So far no fossils confirm this assumption, but it has been proven that chameleons existed in Europe as early as 26 million years ago. The island of Madagascar is still the habitat of several species of chameleons, and some evolutionary theorists suspect that this was the cradle of the chameleons.

Compared to other animal species, the chameleon has some amazing advantages. Its eyes protrude from its head and can be moved independently of one another. A chameleon's field of vision is 90 degrees vertically and 180 degrees horizontally, and this is per eye. With both eyes, the animal reaches a viewing angle of 342 degrees. The perfect all-around view is 360 degrees, so a chameleon's remaining blind spot is just 18 degrees. A phenomenal visual achievement!

The chameleon uses its long tongue, which shoots out of its mouth in a fraction of a tenth of a second, as a weapon. This tongue is not rolled up in the mouth, but rather, like a stretched rubber band, it acts like a catapult. The hyoid bone in a chameleon is flexible and can be moved forward and backward by the muscles. If the hyoid bone is drawn back, the muscles of the tongue tense and the catapult is ready to fire. If the chameleon sees possible prey, the eyes analyze it for size, type, danger, and distance. Then the mouth opens, and at the same time, the muscles of the tongue tighten. The shot is fired, the prey is captured by the tongue, transported into the mouth, and swallowed whole. It gets even more astonishing: the chameleon's tongue does not contain any glue that the prey can stick to. A muscle at the tip of the tongue expands the tissue there into a small, cone-shaped cavity. One-hundredth of a second before the tongue touches the prey, this cavity is pumped out and a vacuum is created sucking the prey onto it. Even more amazing: although the tongue does not contain any glue, the chameleon's saliva is four hundred times more viscid than that of humans. This substance holds the prey in the chameleon's mouth as soon as it is released from the vacuum of the tongue.

Chameleons have to shed their outermost layer of skin throughout their life. New layers are constantly being produced underneath. Each layer must contain the genetic information necessary for the animals to change their color. When chameleons are attacked by bigger enemies, they defend themselves in three different ways: 1) they change their appearance and become invisible to the enemy, 2) they inflate their lungs and plunge into the depths as if they were flying, or 3) they remain motionless and pretend to be dead.

The wonders of evolution are all powerful. In the case of the chameleon, they produce new layers of skin for a lifetime, all of which grow from the inside out and have to change from layer to layer of the corresponding pigmentation. The chameleon's brain combines this information into colors and images on the outer skin. Evolution creates a tongue with no glue, but the same tongue withdraws into a mouth full of super-sticky saliva. Why doesn't the tongue get sticky too? Which evolutionary command created a catapult device with muscles, hyoid bone, and joints in a slow process that lets the tongue shoot out of the mouth in a tenth of a second? And the best part: evolution creates two eyes that can be moved independently of each other; eyes that, in addition, allow an almost all-around viewing angle and provide the brain with a three-dimensional overlay image; eyes that are much more efficient than those of the "crown of creation"—us humans.

The salamander (*Caudata*) is an animal that is similar to the chameleon. Although it cannot adapt its skin to its surroundings, its genes contain the wonderful message of renewal. A part of the body that has been bitten off by an enemy grows back—an ability we humans can only dream of. The fire salamander (*Salamandra salamandra*) can be distinguished from the ordinary salamander by its strong color pattern on the black skin: bright yellow or orange stripes (see Figure 1.13). Like the chameleon, the fire salamander must also shed its skin throughout its lifetime, but its color pattern does not change. Thousands of years ago, people saw the fire salamander as a descendant of the flying, fire-breathing dragon. They also believed that the animal could run through fire unharmed because it was so freezing cold that the fire could not harm it. That's why people threw salamanders into the fire. Around two thousand years ago, the Roman senator and historian Gaius Plinius Secundus (around 23–79 CE) wrote about the fire salamander:

Figure 1.13: The fire salamander (*Salamandra salamandra*)

This animal is so cold that it will extinguish fire when it touches it, just as ice does.[12]

And like bats, the female salamanders have also mastered the trick of keeping the males' sperm in their bodies for months or years. Only when the female wants offspring are the eggs fertilized. This allows the population to be controlled. Why hasn't evolution given us humans the same skills? The problem of overpopulation would be solved.

What actually defines a nation? Genealogy? Religion? Structures? Common goals? Streets? National borders? The army? The leadership? We humans needed 5,000 years of development, from the cave dweller to what we are today: Nations and communities of states. However, in this point too, evolution is ahead of us. State communities, with all the trimmings, have existed for millions of years in ant colonies, including national borders, armies, common goals, streets, dwellings—some for the kings and queens—and also the evils of society, robbery and exploitation.

ANTS: TAKING OVER THE WORLD BENEATH OUR FEET

Ants (*Formicidae*) have existed for 100 million years and there are thirteen thousand different species of them (see Figure 1.14). Their behavior has been thoroughly researched. Wikipedia alone cites sixty-three publications.[13] Ants can form gigantic super colonies made up of billions of animals and spread over thousands of kilometers. One ant colony runs from northwest Spain along the coast to the Italian Riviera—a distance of 5,800 kilometers (more than 3,600 miles)—and is made up of millions of nests. Ant burrows are artificial hills and domes, designed in such a sophisticated way that no water floods their tunnels. The hills can be up to 2 meters (6.5 feet) high and 5 meters (16 feet) in diameter, interspersed with many levels, to which tunnels and paths provide access. On the various floors, there are rooms of different sizes. In these, not only does the queen lays her eggs, but also ants farm mushrooms and care for guests. Since it is warmer in the upper part of the nests than below, the upper floors are used as frost protection areas. The cooler floors serve as cold chambers for the winter rest. In the event of a major flood that also affects people, fire ants hold on to each other to form floating nests until the water retreats and they can find a place to dock—just like the principle of Noah's ark. One type of jet black ant (*Lasius fuliginosus*) even manages to build real structures out of cardboard.

> *They chop up small wood and earth materials and soak this kneaded cardboard substance with honeydew that has been regurgitated from the goiter. This building mass contains 50 percent sugar. Then they grow the fungus Cladosporium myrmecophilum, which gives the nest walls stability through its hyphae (filamentous cell structure typical for fungi). Both living beings live in symbiosis, because the fungus finds optimal food in this way.[14]*

Figure 1.14: The ant (*Formicidae*)

This is true. But why?

In a sensational television documentary, Christina Grätz presented the raids of an aggressive ant species. "I am convinced," she said, "that wood ants are among the strongest, most helpful, intelligent, hardworking and social animals of all."[15] The screen showed how ants scout out their surroundings. When a source of food is found, the animal carries a tiny sample of it home. The ant marks its path with a trail of scents that lure other ants along the same route. The animals very quickly discover the shortest distance from their burrow to the source of food. Because more and more ants follow the route, more and more of the scent sticks to the ground, and an ant trail is created. If the source of food is an injured beetle, it is attacked with formic acid, its antennae are bitten off, and the animal is dragged into the burrow in an ant procession.

Thanks to the arduous work of researchers over many years, we now know the social and often very bellicose behavior of ants.[16] It is breathtaking to find out what ant colonies are capable of and what special properties they have acquired. There are hunters, ranchers, gatherers, gardeners, and slave drivers. Several species of ants carry out regular campaigns in order to enrich themselves. The predatory army ant forms a real line of attack up to 20 meters (more than 60 feet) wide. Any living thing that gets in its way will be rolled over.[17] The slave-making ants of the Amazon rainforest (*Polyergus rufescens*) attack alien nests and kill all animals including the queen. Only the larvae remain. They transport these to their own hill. Each ant drags a larva. In the burrow they fed the larvae until they hatch and then they abuse them as lifelong slaves. This exploitation is not only carried out by the slave-making ant, but also by the predatory blood-red ant (*Formica sanguinea*) and sixteen other species. Did one group learn from the other or is the "predator program" genetic?

For all ant species, the antennae are the most important sensory organs. With them they can smell, analyze air currents and temperature fluctuations, measure and feel humidity, differentiate between light waves and, in particular, communicate. The animals have complex eyes, which in turn are composed of a few hundred individual eyes. Ants with a gender have more eyes than their genderless conspecifics, for example workers. Thus the animals with a clearly defined gender, which are provided with wings, have three forehead eyes, which the workers do not have. All ant species have multiple glands that produce antibiotic substances as well as acid and vapor. The formic acid vapor is deadly for small animals; the liquid acid paralyzes and eats away at the enemy. Some species, such as the fire ant, also have a venomous sting.

All ant species are organized in regular state-like communities, whereby individual colonies can consist of millions of individuals. In every state there is a stubborn caste system: workers, males, and the queen. Evolution (?) arranged it in such a way that no worker can flee—it has no wings. The males are different, as is, especially, the queen.[18] She is not a dictator as we humans imagine, nor an authority in the sense of a monarchy, but simply the mother of the state; she is the only one who gives birth to the offspring and makes community possible. The entire burrow consists of her children. But the queen somehow "knows" whether the state needs more workers or males. Accordingly, she produces different eggs.

Males have wings and leave the ant burrow together with females in swarms. What for? They are to relocate somewhere, make a female the egg-laying queen, and increase the entire ant population. This seems to be the life purpose of all ants. Like in a science fiction film, they could eventually take over planet Earth.

Who hasn't seen an ant trail with little animals that are all transporting a piece of a green leaf (see Figure 1.15)? These are leaf cutter ants (*Acromyrmex octospinosus*), and the green tree leaves that they chop up before they migrate do not serve as food—they end up in the "greenhouse" of the anthill. There, fungi of the genus *Attamyces bromatificus* are grown on the pieces of leaf. The ants apply a thin layer of secretions to the leaf parts. From this, the fungi produce protein-rich threads that serve as a source of food for the ants. The fungi also break down bacteria. A working symbiosis. But why is it so complicated? There is no need to grow fungi to get proteins. Killed small animals also contain the proteins. And why should ants even "know" that they need proteins and receive them from fungi that had been fed before? Which original ant came up with the idea of "fungi cultivation under construction," and how did it pass on its knowledge?

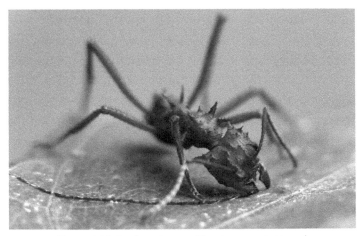

Figure 1.15: Leaf-cutting ant (*Acromyrmex octospinosus*)

The same applies to the species of ants that drag caterpillars into their burrows, feed them there, and protect them in small niches from their predatory conspecifics in order to get their sweet juice. Or for those ant species that enter into a symbiosis with aphids and psylloidea. The ants milk the aphids, and to ensure that those put up with it, they protect them from predators. There are even species of ants that transport aphid eggs into their burrows and let them hibernate there. Wikipedia reports that "wars between different ant colonies have been observed, in which they fought over the dominance of herds of lice."[19] The ant-loving cricket (*Myrmecophilus acervorum*) is also a guest in many ant burrows. The animal belongs to the family of the Ensifera cricket and moves so carefreely in the anthill, as if it belongs to it. Ant-loving crickets are pure parasites. Undisturbed by their host animals, they feed on the eggs, the supplies, and even the larvae of the ants. The biologists suspect that the ant-loving cricket takes on the scent of the ants and is therefore accepted as one of them.

Is all this normal in the world of ants? Is the often brutal way of nest propagation normal? This happens when a queen leaves her den, accompanied by a number of workers. She looks for the nest of relatives, enters, and kills the resident queen. Her offspring will now be raised in the foreign anthill. What is the evolutionary advantage of this? Eggs are laid one way or another: by the previous queen or by the invading queen. And this predatory approach certainly does not match the much-praised social behavior of the ants. Incidentally, young queens mate with many males. A single queen can receive up to 100 million sperm. She can store these in her stomach for years before the eggs are fertilized, just like bats or salamanders. Given the abilities of many animals, we humans could become jealous.

The termites (*Isoptera*) live in a very similar caste system as the ants. Given their physique, one imagines an enlarged, mutated ant. But termites are not ants. They are descended from cockroaches and, in the 30 million years of their existence, have developed 2,900 different subspecies. Each one is feared because termites eat away at woodwork and cause buildings to collapse. In Australia, the annual damage caused by termites is estimated at 100 million dollars, in the US it's 1 billion.

Although the latest genetic research shows that the animals are not related to the ants, they behave in a similar way. Judith Korb from the Albert-Ludwigs-University in Freiburg, Germany, published countless specialist articles on termites together with her colleagues from biology and genetics. Thanks to this work, we now know that a "royal couple" lives in every termite den. This couple stays together for life, which is a strange connection because the male sheds his wings after their wedding flight together. He can't leave anymore. This shedding of the wings is genetically programmed because the break line between body and wing exists in every male. The queen is chained to her den in the same way. Her bottom is so swollen that she can't get away. Who is surprised she lays about 30,000 eggs a day?

Although there is no relationship to the ants, the termites also build the same multistory mounds, or they nest in tree trunks. Some species dig a system of underground passages, others build hill-like towers that can be up to 8 meters (26 feet) high. The termite structures not only exist above ground, but they also reach deep underground. Tunnels as well as ventilation shafts, incubators, galleries, and sleeping chambers are built there. The corridor walls consist either of chewed and hardened wood or earth mixed with saliva. The unassailable outer walls of their buildings are made of a material that's as hard as concrete. Hundreds of thousands of termites and their fungus-colonies produce large amounts of carbon dioxide, and that has to be transported out of the burrow. This is done via ventilation shafts and proper ventilation stacks. The building always has an average temperature of 30 degrees Celsius (86 degrees Fahrenheit).

Workers of both sexes live in the interior of the hills, followed by loads of soldiers, nest builders, food collectors, and brood keepers. And all of them—without exception!—need parasites to exist. Bacteria strains and various unicellular organisms live in the termites' intestines, in such quantities that, according to Wikipedia, they "make up about a third to half the weight of the animal."[20] That is a tremendous amount. For comparison, about 100 trillion bacteria that vegetate in the human intestine weigh a total of around 1 kilogram (2 pounds). Some of the bacteria in termite intestines have enzymes that break down cellulose. This is why the animals digest the wood they eat.

Like ants, some species of termites also grow fungi in their burrows. These, in turn, form small granules that serve as food for the termites. Different species of termites grow different types of fungi—to each its own delicacy so to speak.

The question remains: Why does evolution bring the same knowledge to two different animal species that are not at all related to each other? Both the ants and the termites give birth to entire armies of soldiers that are used exclusively for defense and attack. Now all animals in a termite den have the genetic makeup of sexually mature members—except the soldiers. There is no gender development capability provided in their genes. They are only born to fight; that is their sole purpose in life. Which evolutionary "mind" regulates that?

We humans often speculate about the future and could imagine that using "genetics," we will one day be able to breed clones—genetically identical human beings that are only programmed for a specific job, for example to be a soldier. But hold on: "nature" has been doing just that for millions of years, as is seen by these ant soldiers. And even stranger: the so-called workers in the colony are only there to change their gender when needed. They don't work at all. They do nothing to maintain their colony.

How did the following commands get into the genes of the different ants? You, only build ventilation shafts! You, only produce sperm! You are only attacking—or defending! And so forth. After all, billions of these creatures exist, but each group performs different tasks.

BUT WHO CONTROLS THE ANTS?

One would think that the stable structure of a termite species is impregnable—for small animals anyway—and large animals may rub against it, but the "concrete-fortress" survives it. What it does not survive, however, is an anteater attack. The giant anteater (*Myrmecophaga tridactyla*) is very hairy and reaches a size of 1.40 meters (see Figure 1.16). Genetically speaking, anteaters are related to the sloths. The separation between these two is said to have occurred 58 million years ago. The anteaters developed sickle-like curved toes and sharp claws. With this, they tear open every termite den. Then a very sticky tongue, up to 6 inches long, snaps into the vaults, and thousands of termites (or ants) get stuck on it. The anteater rolls back its tongue and the live prey is scraped off into the esophagus. But the termites fight back. Whole armies of soldiers spray the attacker's tongue, head, and fur with formic acid. That is why the anteater only lingers briefly at the termite mound. His tongue flicks into the cracks of the building two or three times, then he shakes the annoying animals off his body, scrapes them off his face with his paws and trots to the next meal—less than 40 meters (~130 feet) away.

Figure 1.16: The giant anteater (*Myrmecophaga tridactyla*)

One may ask how this "ant eating" started in the first place. The anteater (or, more accurately, one of its ancestors) stuck its tongue in an ant or termite den. The residents resisted. Soldiers sprayed acid on its tongue and crawled into the anteater's eyes and ears. Usually an animal runs away during such attacks, shakes itself, cleans itself up and never comes back. Why didn't the anteater's ancestor do this?

Evolution gave the anteater a long, sticky tongue and a very small mouth opening. In addition, it received that cone-shaped head, which becomes ever pointier toward the mouth, and the ability to stand on its hind legs—all qualities it needs for its special type of hunt. Why do similarly built mammals not also hunt ants and termites? The raccoon (*Procyon lotor*), for example, has an equally pointed head, it can just as easily stand on its hind paws, and its claws are just as capable of ripping open an ant burrow. But it is not interested in ants and termites. This fabulous evolution developed other skills. Raccoons can stand in cold water for hours and climb down a tree head first. Nothing special? Oh, but yes! Normal animals climb backward down the tree. Raccoons, on the other hand, are able to twist their hind paws so that they point in the opposite direction. In addition, the animals are classified as having above-average intelligence. In captivity, they solved amazing tasks and remembered them years later.[21] And they only got their German name, *Waschbär*, because they swirl their food thoroughly in water before consuming it. In Spanish, they are called *mapache*, which in turn comes from the Aztec word *mapachitli* meaning: "The one who checks everything with his hands."[22]

Except ants and termites are not on their menu.

The so-called scaly anteaters or pangolins (*Pholidota*) are different. For centuries, it was believed that this species was related to the anteaters and armadillos because they lick up ants and termites just like anteaters do. But modern science came to a different conclusion. Wikipedia writes:

> *From the mid-1980s, molecular genetic studies showed that pangolins are more closely related to carnivora.*[23]

The first forms of pangolins are said to have developed as early as 47 million years ago. Their behavior, which is similar to that of the anteaters, is classified as an adaptation. Adapted to whom?

The animal's body is covered with scales. These overlap like roof tiles and have sharp edges. If such an animal is attacked, it will curl up into a ball. Then the individual scales stick out from the body like teeth or half-open pine cones—hence it's also called "pine cone animal." In addition, pangolins can produce a disgusting stench that drives any attacker away quickly. Although not related, the pangolin is just as keen on ants and termites as the anteater. The animal locates its prey with its nose. Then it breaks the burrow open with its sharp claws. As with the anteater, the animal's palate hides its tongue, which is provided with a bundle of muscles. It can be stretched up to 25 centimeters (almost 10 inches). Termites or ants stick to it and are scraped off and transported into the esophagus. Then the stomach takes care of chopping up the prey because pangolins have no teeth. To prevent foreign soldiers from entering their bodies and spraying venoms there, they seal their eyes, ears, and nostrils while they are eating.[24]

Since pangolins are not related to the anteater, there can be no common genetic characteristics between them. But then why do both species hunt in the same way? Why are they just as keen on ants and termites, and why do they both use the same method—the tongue—to transport their victims into their digestive tract? Has evolution played a double game? Does that also apply to the strange worm-like creatures called centipedes or millipedes?

MANY LEGS, MANY PERMUTATIONS

Both exist: centipedes and millipedes. With the centipedes (*Chilopoda*; see Figure 1.17), between 15 and 191 pairs of legs were counted, depending on the species, whereas

Figure 1.17: Centipede (*Chilopoda*) feeding on a
slender blind snake (*Leptotyphlopida*)

millipedes (*Myriapoda*) have up to 750. The number of
legs is uneven in all centipedes. Why is not clear. The first
pair of legs turned into a single venomous claw. Some
of the species produce highly venomous hydrogen cyanide, others greasy substances with which they glue rivals'
mouthparts together. The variety of evolution is limitless.
"At night, they go on long, extensive forays as active hunters who pursue their prey and overwhelm them at lightning speed. In doing so, they shoot forward, like snakes,"
Wikipedia reports.[25] There are said to be at least 3,000 different varieties of millipedes, perhaps even up to 8,000.
Among them are the *Ethmostigmus rubripes* with a length
of 16 centimeters (~6 inches), the *Cormocephalus*, measur-
ing 25 centimeters (close to 10 inches), and the Amazo-
nian giant millipede, which can be over 30 centimeters
(11 inches) long. The animals are carnivores that not only
dismember tarantulas, scorpions and insects, but also

"smaller vertebrates such as bats, birds, snakes, frogs and small lizards."[26] There are venom glands in their mouths, and they even destroy the chitin shell of their victims with their claws. There are centipedes and millipedes with and without eyes. They are all dangerous, and they locate their victims using various sensors.

Evolution was not satisfied with ten or twenty varieties of the same species; it developed thousands upon thousands, new mutations every time, whereby—according to most schools of thought—most mutations lead to a negative result that do not help the animal improve any further.

Now, one should know that many species of these polypods produce hydrocyanic acid (hydrogen cyanide). This also occurs in nature in microscopic amounts, for example in the kernels 'of stone fruit (apricot, peach, cherry). But just one milligram of hydrogen cyanide per kilogram of body weight is absolutely fatal. Millions of people were murdered with hydrogen cyanide in the Nazi extermination camps during World War II, and until 1999, death sentences in the US were carried out with hydrogen cyanide. What is evolution doing? A small, worm-like animal with innumerable legs produces the same hydrogen cyanide and thus poisons its victims. The amount of hydrocyanic acid produced in this animal is small, but the multiped is also very small. How did the millipedes develop their own immunity to the highly toxic level?

A splash of this acid that missed its mark would seemingly affect the multiped too. In addition: the killed victim, when it is finally devoured, also contains hydrogen cyanide. The situation is similar to that of the bombardier beetle. If the dosage of hydrogen cyanide given off by the multiped was not highly toxic, its victims would not die immediately. How could the venom develop slowly in the body over time? This is a question that ultimately affects all animal species that use venom.

EXPLORING VENOMOUS SNAKES

Around one hundred thousand people die each year from snake poisoning. Of the 3,200 species of snakes that crawl around on our globe, around half are venomous, with the reptiles producing completely different venoms. Some only attack the cells of the heart muscles and lead to cardiac arrest. Others cause muscle paralysis, still others mainly destroy the kidneys, and other toxins cause internal bleeding.[27] In toxicology, the amount that can be lethal for various lifeforms is known by the term *lethal dose (LD)*. For example, the LD of potassium cyanide for an adult person is 140 milligrams. The lethal dose is LD75, the definitely lethal is LD99 and the absolutely lethal is LD100. The lethal amount is relative and depends on the size and other circumstances of the lifeform that is bitten. The most venomous of all snakes is the inland taipan (*Oxyuranus microlepidotus*) in Australia. Its venom is 50 times stronger than that of an Indian cobra. Its LD for humans is 100 milligrams. The dose of venom that an Australian inland taipan injects into its victim per bite would be enough to kill 230 people.

Each snake possesses different venoms.[28] This is even true for sea snakes (Laticaudinae), even though one might think they didn't need any venom living in the water (see Figure 1.18).

But evolution did not stop at one species of sea snake: fifty-six of them exist, including the annulated sea snake

Figure 1.18: Common, or blue lipped, sea krait (*Laticauda laticaudata*)

(*Hydrophis cyanocinctus*) with a length of 2.5 meters (~8 feet) or the yellow sea snake (*Hydrophis spiralis*) with 2.75 meters (9 feet). The reptiles are related to each other, because they all have a laterally flattened tail and a gland under the tongue with which they absorb excess salt. The animals must have moved from land into water tens of millions of years ago, as their oversized lungs suggest. That allows them to dive for up to 2 hours and to a depth of 180 meters (almost 600 feet). Originally, researchers thought that sea snakes would only eat fish. In the meantime, it has been observed how they swallow prey twice their size. The salivary glands of sea snakes mutated into venom glands, with the venom being injected into their prey through their teeth. The most venomous of its kind is the Dubois' sea snake (*Aipysurus duboisii*). In the *MeerwasserLexikon* (Seawater Dictionary) it is described as "very venomous."[29]

Even an inflated animal like the puffer fish (*Tetraodontidae)* is poisonous (see Image 8 in the color insert).

A dwarf puffer fish can grow to a size of just 3 centimeters (~1 inch); its brother, the giant puffer fish or the starry puffer *(Arothron stellatus)*, grows to a length of over 1 meter (3 feet). In the event of danger, the animals pump themselves full of water, with the spines erect on their skin that have barbs at the tip. This makes it impossible for the predators to take their victims by the throat. In addition, the internal organs of the puffer fish are toxic. Tetrodotoxin (TTX)-type venom is a neurotoxin that completely paralyzes the victim's muscles. They can neither move nor make a sound, but they remain fully conscious. A slow horrific death occurs from respiratory failure.

DIFFERENT ANIMALS, DIFFERENT VENOMS

Scorpions (*Scorpiones*) are assigned to the genus of arachnids. To date, believe it or not, we know of 2,350 different variants. Little is known about their evolution. It is believed that they are originally descended from a marine crustacean that crawled ashore around 400 million years ago. The abdomen of the animal consists of chitin rings, the last of which forms a tail with a poisonous sting.[30] Inside the chitin rings there is a poison gland. When hunting, the scorpion first uses its grasping claws to grab its prey, then, a fraction of a second later, the venomous sting shoots forward through the tail from the abdomen (see Figure 1.19). As with a hypodermic needle, the venom is injected into the enemy, which is then killed and dismembered. The venom is highly effective especially since the scorpion can use two different venom types: one component kills arthropods such as beetles and millipedes; the second it uses only for defense. So it affects opponents who are not to be eaten, but only to be fought off. In the scorpions, as in other animals, the venomous mixtures are completely different.[31] The LD values are between 50 and 100. The yellow deathstalker scorpion (*Leiurus quinquestriatus*) is considered to be the most dangerous scorpion on Earth. It lives in arid desert areas. Its venom is made up of six chemicals: chlorotoxin, charybdotoxin a and b, and agitoxin 1, 2 and 3. But, strangely enough, some animals are immune to the venoms of scorpions.

Figure 1.19: Scorpion (*Alacranes gueros*)

Even more deadly is the poison of the golden poison frog (*Phyllobates terribilis*), also known as poison dart frogs, which is considered to be the most poisonous frog on our planet. It lives mainly on the Pacific coast of South America but is also found in the primeval forests of the Amazon. Little is known about its ancestry. Even touching its bright yellow skin is fatal (Figure 1.20). The indigenous people learned quickly how the poison of this animal could be used as a weapon. They used prey to lure the frog into traps, then arrowheads were rubbed against its skin and left to dry. They also used these poisoned arrows on the Spanish conquerors, who neither knew of poisoned arrows nor possessed any means of defense.

Amazingly captured poison dart frogs lose their poison, and offspring develop as nonpoisonous frogs.

Figure 1.20: Golden poison frog (*Phyllobates terribilis*)

All of the venoms and poisons covered so far arise in the bodies of animals: snakes, spiders, scorpions, millipedes, fish, and so on. How does this come about? One would like to imagine that, in its distress, a frightened, attacked animal might have worshiped some ominous "spirit of evolution" and urgently asked for a weapon, because otherwise it would not have been able to survive. A crazy thought? How does murderous venom develop in a lifeform that, according to our understanding, has no brain with which any commands can be formulated? Does a jellyfish have a brain or do the muscles of the animal react automatically? The box jellyfish (*Cubozoa*) is such an impossible thing (see Image 10 in the color insert). It is one of the most feared species of jellyfish, the so-called sea wasps. Children touched by the tentacles of this jellyfish die within minutes. A tentacle of the genus *Chironex fleckeri* can reach 3 meters (almost 10 feet) in length. A full-size specimen of this monster has sixty such tentacles. Around one thousand so-called explosive cells (*nematocysts*) hang from every centimeter of these tentacles. When touched, they shoot into the victim's skin like harpoons in a flash. In his witty and humorous book *Die fabelhafte Welt der fiesen Tiere* (The Fabulous World of Nasty Animals), biologist Frank Nischk describes it as follows:

> *The explosive cell literally explodes, and some of these cells at the same time hurl a sharp spine, known as a stylet, at the external barrier, which, because the acceleration is so enormous, can even penetrate the shells of shellfish or the scales of fish.*[32]

The author refers to pictures that were taken with ultra-high-speed cameras and prove that "only three thousandths of a second pass from the first contact until the poison reaches the victim."[33]

A FEW INTERMEDIARY QUESTIONS

The millipede can literally "stuff" its prey's mouth with a bite. This is due to a fluid it uses that can glue the prey's mouth shut like a rubber seal. Why does the fluid not harden at the edges of its own mouth? How is this fluid supposed to have condensed more and more over millions of years, but only outside of its own body?

The millipede produces various poisons, including hydrogen cyanide. Why do animals of the *same* species develop *different* poisons? How did immunity to one's own poison arise? Slowly? Even "a little poison" must have been fatal for the prey in the early stages.

At least 1,400 species of snakes, all related to one another, produce completely different mixtures of toxins.

Sometimes the toxins cause cardiac arrest in their victims, cause the blood to clot, destroy the kidneys, or paralyze the muscles throughout the body. The answer to the question of how all of this came about—"It just happened that way over millions of years"—is not a scientific answer. Shouldn't the snake venom have the same composition in all species?

Of 56 species of sea snakes, some developed venoms. Their salivary glands turned into venom glands. This change cannot have happened in a short time, because cannulas had to be formed inside the teeth to the transport of poison. So the development went on for millions of years. In any case, rapid development seems to be impossible, otherwise the snake would have poisoned itself. And in what process, lasting millions of years, did the mixture of poisons, the cocktail, come about?

The tiger shark is immune to the deadly venom of the most poisonous of all sea snakes, *Aipysurus duboisii*. How did his immunity come about? Are the animals supposed to have cuddled with each other millions of years ago and gotten used to each other?

The venom of the poison dart frog is distributed over the skin of its body. But an attacker does not know that. First of all, it bites down. If it spits out the frog again or lets it go, it will be too late. How would attackers know that the frog is deadly? Why does the poison dart frog lose its venom protection in captivity? What chemical processes take place in its body then? Why are the poison dart frogs born in captivity nonvenomous? Shouldn't the frog, especially in captivity, defend itself against its captors and keep its venom?

The jellyfish of the sea wasp genus developed 3-meter-long (10-foot-long) tentacles that are equipped with hundreds of thousands of highly poisonous explosive cells. In a split second, these are shot into the victim's skin. Whatever comes within 3 meters (10 feet) of a sea wasp is done. Shouldn't this monster be invulnerable? How should you imagine a slow developmental process that enables a shot to be released in one hundredth of a second? If the sea wasp's defense were to take place automatically, comparable to radar that triggers defense systems, then the sea wasp itself would not be able to approach a victim, let alone any animals that could have developed immunity to it.

THE FACT IS—IT'S ALL REALLY JUST LIKE THAT

Why? Why? Why? Year after year, in November, thousands of tourists gather on Christmas Island, which is located in the Indian Ocean and, in political terms, belongs to Australia. They are there to witness a unique spectacle: hundreds of thousands of Christmas Island red crabs (*Gecarcoidea natalis*) run a 5 -to-7-kilometer (3-to-4-mile) stretch from the forests to the sea. While making their dash, they also climb over obstacles such as tree trunks or small walls. When they arrive at the coast, they spray each other with salt water.

The males then dig a small pit and receive the females there to mate, which can last up to 20 minutes—sex on the beach. After the sexual act, the females run into the water and press their eggs out of their bodies. Even if millions of these eggs are eaten by fish, several hundred thousand crabs hatch. The crabs then promptly wander back into the woods. The following year the game repeats itself, as it has for millions of years. By the way, the human population of Christmas Island is around 1,500—the number of crabs is estimated at 45–80 million.[34]

There are around 6,800 species of crab worldwide. Among them is the short-tailed crab of the infraorder *Brachyura* (see Figure 1.21). It spends its larval stage in the sea. Only after metamorphosis does it swim and then run back to its place of birth as a miniature crab. Where the animal, while drifting in the endless ocean, receives the geographic coordinates to reach its destination from, remains a mystery. According to the relevant science, crabs are descended from a common maritime ancestor that existed around 4 million years ago.[35] Why does the short-tailed crab have a navigation system that guides it precisely to its place of birth, but countless other crab species do not? Even though they have a common ancestor?

Figure 1.21: Short-tailed crab (*Brachyura*)

The next animal with highly amazing properties is the mantis shrimp (*Stomatopoda*). There are 400 different species of mantis shrimp, and German marine researcher Helmut Debelius differentiates between them based on their hunting behavior—whether they are "spearers" or "smashers."[36] The spearer's legs are pointed. With these, it literally spears its prey. The smasher is different: with help from its strong muscles, its tentacles shoot forward at 82 kilometers (roughly 50 miles) per hour. This is, according to Wikipedia, "one of the fastest movements any animal carries out. The impact is similar to that of a pistol bullet."[37] The impact is so powerful that it splinters the shell of the shrimp's prey. The victims are stunned immediately. At the same time, the mantis shrimp creates a bang and a flash of light.

The weapons of the mantis shrimp are reminiscent of those of the pistol shrimp. But the two animals have different ancestors.

We respectfully marvel at "the wonders of nature" and imagine that this "nature" must be something like a spirit that makes miracles possible. And it teems with miracles that will one day be scientifically explained. But in the year 2021, there are no answers for innumerable wonders of nature, and faith alone isn't a satisfying explanation. So I want to ask more questions.

THE STRANGENESS OF BIRD MIGRATION

There are around thirty thousand species of flies worldwide, some of which live for a few weeks, others for just a day. All of them have mastered the technique of crawling around on mirror-smooth walls or ceilings. How do the animals manage that without miniature suction cups? Hair grows between their legs that end in microscopic ovals. Inside each is a mini film of a sticky substance that allows the flies to adhere to very difficult surfaces. The substance is chemically so finely dosed that the fly can walk without getting stuck itself. Bravo to evolution, which has produced this miracle trillions of times every day for hundreds of millions of years. Flawless. A startled fly takes off instantly from a still position, flies in all possible directions and lands just as perfectly. Considering all the aircraft *we* humans have developed—they are no match to what flies can do.

Why do around 50 billion birds undergo annually the agony of a seemingly endless flight to other regions? Including 5 billion that fly between Europe and Africa? The common and mostly correct answer is: because of the climate. It gets too cold for the animals in the north during the winter, and they run out of food. So they head south. But this explanation only applies to some of the bird species. Others leave their breeding areas regardless of climatic conditions year after year.

A single bird equipped with a transmitter was observed flying non-stop from Alaska to New Zealand.[38] Distance: 11,500 kilometers (over 7,000 miles)—an incredibly long route. All the bird saw below it was the sea, day and night. The small animal couldn't have orientated itself by the stars. They're not there during the day, and on the route between Alaska and New Zealand, the star constellations change from those of the northern to those of the southern hemisphere.

The magnetic field as an orientation aid is only a partial solution. A kind of "magnetic receiver" was discovered in the bodies of various bird species; with robins it was in their right eye, and with pigeons it was in the skin of their beak. But the magnetic field changes on the way between the northern and southern hemispheres.

How about "flying by sight?" In fact, research has shown that birds can remember the route of highways or the lighting of large cities. The suggestion is still unsatisfactory, however, because most bird flights take place at night. The huge swarms spend their days on the ground. Only larger animals such as cranes or storks travel during the day.[39] Ornithologist Peter Berthold, who has spent his life studying the behavior of birds, is convinced that both the direction of flight and the duration of flight are innate in birds.[40] However, it must be remembered that both of these differ totally from one another in the different bird species. Some flocks of birds fly from Europe via France, Spain, Tunisia, and the equator to South Africa, others take travel through Italy and the Mediterranean to the south, and still others choose to travel over Greece, Turkey, Lebanon, and Egypt for their flight. Migratory birds even cross the Himalayas. Birds have been observed reaching an altitude of 10,000 meters (almost 33,000 feet). Up there the temperatures are 40 to 50 degrees Celsius (–40 to –58 Fahrenheit) below zero.

Obviously these birds are taking different routes to different destinations. So no primordial command is anchored in the genes of the birds, otherwise they would have to fly the same route year after year. Therefore the situation cannot be compared with the genetic message of salmon or eels. And why do many thousands of birds follow their leading birds? How are the commands passed on to the whole flock during the flight? And which great-great-ancestor started it?

The Arctic tern *(Sterna paradisaea)*, an animal with a fiery red bill and red legs, covers (at least) 100,000 kilometers (more than 62,000 miles) every year (see Figure 1.22). The bird, which is just 35 centimeters (just shy of 14 inches) tall and only weighs 100 grams (3.5 ounces), breeds in the Arctic and spends winters in the Antarctic. The tern's breeding ground and the place where it hibernates are at the other end of the Earth. Scientists have attached very light transmitters to these birds that deliver GPS data that prove that the birds cover this gigantic distance in stages of around 700 kilometers (more than 400 miles) a day. The birds also chose different routes from one pole to the other: they did not take the most direct route, but detoured across the entire Indian Ocean, via South Africa and Australia.

It is not known why the tern flies from pole to pole and back every year. Luca Börger from the University of Swansea in Wales, Great Britain, suspects that the bird

Figure 1.22: Arctic tern (*Sterna paradisaea*)

may be an extreme sun worshiper who loathes the dark. After all, it is "the animal that sees the most sun of all."[41]

And what about a crane migration? If you haven't seen one, take a look at Figure 1.23. A structure appears in the sky that looks like a giant wedge or arrow, but sometimes it also looks like a row of objects flying at an angle. These are Eurasian cranes (*Grus grus*). The animals start their flight from the ground with quick steps and stretched necks. Then they launch upward with mighty wing beats until they reach their sedge. Cranes are long-haul fliers. They can cover distances of up to 2,000 kilometers (more than 1200 miles) nonstop, often while gliding. As was done with geese, courageous researchers have also participated in crane flights using appropriate technology.[42]

Usually, however, the longer routes are bridged in stages. Cranes are native to northeast Europe and Northern Asia. And, like other bird species, they do not head

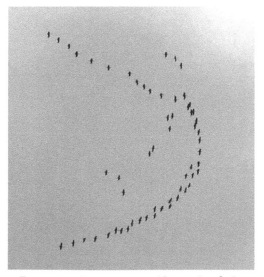

Figure 1.23: Eurasian cranes (*Grus grus*) in flight

for a common destination country. The fly several different routes—sometimes they fly via Italy and Sicily to Tunisia, then via Cyprus, Israel, and Egypt to the Red Sea; or via Western Siberia and Central Asia to Pakistan and India. The flocks can contain 20,000 animals.

The same applies to the migration of the white storks (*Ciconiidae*, see Figure 1.24). These long-haul pilots travel up to 20,000 kilometers (more than 12,000 miles) to reach their winter quarters in Africa, from which they fly back to

Figure 1.24: The white stork (*Ciconia ciconia*)

their breeding grounds next spring. Experts distinguish nineteen different species that live on all continents with the exception of Antarctica. Fossils of the animals prove these storks, existence during the Oligocene—around 30 million years ago. Like other migratory birds, white storks reach their destination via various routes; sometimes via Spain and North Africa to the Sahel, or via Italy and Tunisia to Senegal. The amazing thing is that most stork species are not migratory. They stay sedentary all year round.

On the route between Évora and Lisbon, Portugal, you drive past several old high-voltage pylons. You can see a stork's nest on each. In the Swiss village of Altreu, in the canton of Solothurn, you can admire several stork families in their nests on the chimneys of farmhouses (see Figures 1.25 and 1.26). In fact, Altreu has been designated the official "stork village," protected by authorities since 1948. The storks in Portugal and those in Switzerland are migratory birds. Every year, after a 1,700-kilometer (1,000+-mile) flight, they return precisely to their pole near Lisbon or their house chimney in Altreu, Switzerland. An internal navigation system, which the storks can perhaps tap into using earthly magnetic fields, may guide them to their destination country after thousands of kilometers, but not precisely enough to get them to the same high-voltage pylons in Portugal or the chimney of the same farmhouse in Switzerland. Just think: Switzerland and Portugal are just two of many destinations. However, storks fly to thousands of different locations around the world and find them precisely every year. This also applies to the descendants of the stork parents: the young play the same game. How exactly do they inherit this navigation system? It cannot be a message that was entered into the stork's genome by some archaeopteryx, a so-called primordial bird-like dinosaur.

Figures 1.25 and 1.26: Storks nesting on rooftops in Altreu, Switzerland

THE IMPOSSIBLE OCTOPUS

A creature with downright uncanny abilities is the octopus, popularly known as squid or kraken (see Image 11 in the color insert). Octopuses belong to the group of

Coleoidea. The original ancestor of this species is unknown, but its evolution from a worm-like being must have started around 700 million years ago. This cannot be proven. Just like the chameleon, an octopus can adapt its color to its surroundings in a fraction of a second. Laboratory experiments showed that the octopus changed color 177 times in an hour.[43] What incredible computing power! The latest research conducted by Stanford University professors Benjamin Burford and Bruce Robinson at the Monterey Bay Aquarium in Monterey, California, showed something astonishing: octopuses communicate with one another using their colors. "The colored signs on the skin were so detailed and complex that the animals would be actually able to convey precise messages."[44] To this day, Hawaiians say that the octopus is the survivor of a world that completely vanished.

Octopuses puzzle researchers. After all, with the octopus, you don't know where the brain begins. A network of neurons extends over the entire body, including all arms." Philosopher and octopus researcher Peter Smith compares this brain to the body's own Internet with 500 million nerve cells. In addition to their odd "brain," octopuses have three hearts that pump bluish blood through their bodies. While octopuses have eight arms, squids have eight arms and two longer tentacles. In the giant Pacific octopus, these arms reach a span of up to 7 meters (22 feet). Each arm is equipped with innumerable suction cups, and each suction cup has ten thousand neurons. Octopuses can use their arms to analyze chemicals, light, and many other things. They can move quickly in any direction from a resting position. Also, since squids and octopuses have no bones, they can give their bodies a wide variety of shapes. There are known cases in which octopuses have crawled from their aquariums into the public sewer system and from there into the sea. Others have found a way from a neighboring aquarium back to their own pool. They manage these feats because they can slip through narrow cracks and crevices. Marine biologist Joshua Rosenthal reports on experiments in which squids opened screw caps, escaped from aquariums, and were able to clearly differentiate between several things and several people as well.[45]

The fertilization and the lifespan of octopuses are also a mystery. During mating, the male slowly pushes its third, left arm into the female's mantle cavity. The process can take up to an hour. At the tip of the arm is a capsule filled with sperm, which bursts explosively into the body of the female and releases its contents. Then the female lays hundreds of thousands of eggs that are attached to small stalks. A few weeks after mating, the female dies. The young hatch without parents; no one is around to look after the brood. Nobody teaches them how to hunt, what to eat, that seals are dangerous, or how to catch shrimp. How do they know what is healthy? What makes them sick? How to deal with enemies? When to use their ink-like dye? In contrast to birds and mammals, octopuses have no parent-child relationship, so how is information passed on?

Most species of octopuses are nonvenomous with the exception of one: the blue-ringed octopus. The chemical composition of their poison corresponds to that of the puffer fish, although the animals are not related. How does the blue-ringed octopus know about its venom and how to use it? It learned nothing from its parents. And it gets even more sinister: this octopus can change its DNA. How does it do that?

Today, every high school student knows what DNA (deoxyribonucleic acid) is. Each and every human and animal cell contains its genetic code—genetic information. Nowadays, anyone who knows this genetic code can tamper with any living being—you just have to know how. Geneticists know. Doing so requires specialist knowledge, laboratories, and high-resolution microscopes. Such complex laboratory work used to take months and be quite expensive. That is, until six years ago, when a brilliantly clever lady at the Francis Crick Institute in London discovered a simpler method for changing the genetic makeup: the CRISPR process. International journals and newspapers reported about this sensation.[46]

In this CRISPR process, "molecular scissors" are used to cut the DNA strand exactly at the desired location. This has recently been called genome editing. The German news magazine *Der Spiegel* wrote "CRISPR can be done easily."[47] But what does that have to do with the squids?

Biologists who study the genetic code of these animals were astonished to find that the squid changes its code on its own—in other words, it rewrites the information. According to this, a CRISPR process takes place in the squid, without electron microscopes, and without sterile laboratories and supermicroscopic tools. After the first findings about the changes in the genome of squids had perplexed the scientists, a team led by geneticist Isabel Vallecillo-Viejo from the University of Puerto Rico and the Biological Research Center in Woods Hole, Massachusetts, investigated the issue. The results are clear: octopuses can rewrite their own genetic code.[48]

To understand this, you first have to know that the four basic building blocks of the genetic code are the chemicals adenine, guanine, cytosine, and thymine. They form the double helix—the twisted ladder—of the DNA. These four building blocks stick together in a very specific order. To change this base sequence, humans need our sophisticated laboratories, and even then, the task is not an arbitrary exchange of a base, that is, some random mutation. Geneticists need to know *which* base is to be replaced by *which other one*. Otherwise, there will be deformities and chaos in the code and therefore also in the body of the lifeform in question. The squid knows *which* bases have to be put *where*; something that was previously considered unthinkable.

All of this is difficult to reconcile with the usual explanations of the theory of evolution. Is there any "spirit" smiling in the strange brain cells of the octopus, which are, after all, scattered throughout its body? Do animals consider us humans to be the ones who are in cages? Do they pass on their knowledge to their offspring in a way that previously seemed unthinkable?

CARNIVOROUS PLANTS AND OTHER CREATURES: DO THEY THINK?

Similar questions arise with carnivorous plants (*carnivores*). If you used to think that such plants were a playful peculiarity of nature with maybe three or four variations, let me disabuse you of that notion. A thousand different species, divided into seventeen genera, are documented. And each species has developed its own way of catching prey. For instance, there are those that lure their victims into simple pits. Once inside, there is no escape. With the so-called corkscrew plants (*Genlisea*), a genus of carnivorous plants, it gets more complicated. Insects and beetles are attracted by fragrances. The prey crawls into a can-shaped cavity, the way back out is blocked by stiff hair. Still other plants developed a sticky, greasy substance. The prey gets stuck on it, and the more it tries to free itself, the more hopeless the situation becomes—comparable to an insect in a spider's web. Some plants have even managed to set up traps under water. The leaves form a small cavity in which negative pressure builds up. With every touch, this cavity opens and the prey is sucked in because of the briefly working suction effect. From an evolutionary point of view, this method is difficult to understand unless the plant is able to plan this.

Planning requires thinking, and with the plant called Venus flytrap (*Dionaea muscipula*), fabulous evolution must have planned all this in advance. This plant has trapping leaves, as the experts call them, and these leaves are equipped with pointed bristles.[49] They have it all. They have joints—that means they can be moved—and they react via sensors of some kind. As with all carnivorous plants, the prey is attracted by honey and a red color (see Image 12 in the color insert). If the victim touches one of the bristles, nothing happens for 20 seconds. Only when a second bristle is touched does the trapping leaf collapse in a fraction of a millisecond. This corresponds to "one of the fastest known movements in the plant kingdom."[50] Meager-looking prey is allowed to escape through the spaces between the bristles.

What gets stuck in the trapping leaf is first examined for its usability. If the plant gains the impression that it is *not worth it*, it opens its trap again. However, if the prey is usable, the leaf trap is literally sealed. Then the plant releases some chemicals to dissolve the prey.

What is happening here is planning for the future. So what? All animals plan for the future. A spider builds its web—future planning. Ants build up food supplies—future planning. The dog buries his bones—planning for the future. But the Venus flytrap *is not an animal.* The plant cannot run around chasing its food. Evolution does not create bristles with joints. The Venus flytrap has no counting mechanism, and no clock. But somehow it counts down 20 seconds before it snaps shut, and it only does so after the prey has touched at least two bristles. Strictly speaking, this is counterproductive because the victim could get away within these 20 seconds. Evolution may well produce leaves that close slowly—but not within a millisecond. And only after a second bristle is moved does the trap snap shut. How did the signal get from the bristles to the joints of the leaves in a millisecond? Which lines of thought are necessary to analyze the size and usability of the victim and then—after a waiting period of 20 seconds—to finally seal or open the trap? Aren't we expecting a little bit too much from evolution?

Had enough of the curiosities? Just wait, it will get even better!

On March 18, 2020, broadcaster ARTE showed a documentary by Jacques Mitsch, which was announced as follows:

> *It's neither animal nor plant, but—a blob. This slimy superorganism puts everything into question that humans think they know about intelligent life. The fascinating single-celled organism is almost immortal, has an insatiable appetite, can solve complex problems and shows amazing learning and communication skills.*[51]

Come again? Neither animal nor plant and still intelligent? That doesn't make any sense. I followed this blob and learned to be amazed again.

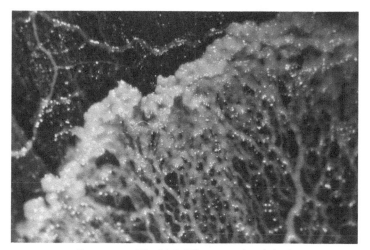

Figure 1.27: The blob (*Physarum polycephalum*)

The blob *(Physarum polycephalum;* see Image 13 in the color insert) is a species of slime mold. Strictly speaking, "slime molds are not real fungi at all. Just as they are not plants or animals."[52] The "thing" consists of a single cell with millions of cell nuclei and can be several square meters in size. The blob owes its name to a science fiction film from the 1950s: Called, not surprisingly, *The Blob.* In this film, planet Earth is attacked by a jelly-like monster from outer space. In fact, the real blob has several of the characteristics of the blob in this science fiction horror movie.

The blob has neither eyes nor ears, it has no nose, no mouth, no legs. It cannot taste or see, digest or feel. It has no nervous system and no brain either. And yet it solves complex problems and develops sophisticated strategies—there are lots of contradictions. Behavioral researcher Audrey Dussutour from the University of Toulouse, France, subjected "the thing" to years of tests and reported how the blob broke out of the test tubes at night and looked for food. Madame Dussutour mixed puddings with various egg whites and sweeteners to see what the blob would prefer. Soon she also noticed that the thing was keen on oatmeal. But how was that supposed to work when the blob had no stomach and no digestive tract? It would just crawl over the food and the food would disappear. The thing devoured bacteria, yeast, fungi, puddings, and, of course, oatmeal. Its speed of movement was 1 centimeter (a third of an inch) per hour. When it was going after food—that is, when it was hungry—the speed was 4 centimeters (an inch and a half) per hour.

Currently, the thing keeps several scientists quite busy. Japanese professors Toshiyuki Nakagaki and Atsushi Tero from Hokkaidō University are masters of the blob. The results of their test series are breathtaking. Professor Nakagaki built a U-shaped barrier between the blob and its food. The blob couldn't see the food source—it has no eyes. Smelling it was just as impossible—the blob has no sense of smell. Nevertheless, the creeping mass found the source of food by the shortest route. Several obstacles were set up, even a maze with a piece of food hidden somewhere. No problem for the blob. It always chose the shortest route. Soon the Japanese scientists realized that the blob didn't like salt. If there was a thin layer of salt on the way to the food source, the blob crawled past it. Small side walls were put up so that the blob had to crawl over them on its way to the food. The blob learned and gradually got used to the salt. Then the blob was cut into several pieces. Offshoots formed, and they too were used to salt, but those blob parts that had not experienced the salt experiment did not.[53] In other words, it seemed as if the blob had inherited what it had learned. How? The thing has neither a brain nor a sexual apparatus.

In Germany, Hans-Günther Döbereiner from the University of Bremen researches the phenomenon of the blob. He is a specialist in the new research area "Evolution of basal cognition in single-celled organisms." Together with his doctoral student Jonghyun Lee, he discovered currents in the "veins" of the blob. They call it "protoplasm." The word comes from the Greek and means: *proton* = the first; *plas*ma = the formed. The blob has vein-shaped structures that slowly contract and then relax again. This explains its movement and also the flow of "protoplasm" in the "veins." But there is no reason why the blob moves in every direction—there is no "front" or "back"—and there is no reason why every part, even if a hundred pieces are cut from a blob, has the same information. At the University of Florence, Stefano Mancuso and František Baluška studied the riddles of the mind without a brain. How did the blob store information without brain cells? And at the Tufts University School of Engineering, scientists working with Michael Levin are trying to find out how the blob communicates.[54]

The blob acts like an ignition spark for many areas of science. Why can the thing do the things it does? Why is a single-celled organism able to learn without a brain? Scholars from several countries exchanged blob pieces with one another. The French Dussutour made a funny statement: there were blob samples from various countries on the laboratory table, and the researcher had just poured some oatmeal from an American manufacturer and some from a French manufacturer onto the table. The American blob sample aimed precisely for the American flakes and ignored the others.

We know that ants leave a scent trail behind to find their way. The blob leaves a microscopic trail of slime behind and it does not cross it a second time unless it is forced to do so by obstacles. The blob knows that it has already been there. If two blobs are put there together, they exchange their information. How can an organism that only consists of the same cell move, absorb information, analyze it, and pass it on to its offshoots? Incidentally, "the thing" also seems to be immortal. If there is no food, the blob dries out, but a little moisture, for example by brushing it with some water, is enough to bring it back to life.

The blob has become an attraction in the zoological garden of Paris. The plan is to shoot a piece of blob into space in order to expose it to weightlessness, radiation and even vacuum. I'm not so sure this is a good idea. Maybe space is the blob's original home, and it is multiplying out there billions of times and invading Earth like a virus.

Whatever "that thing" is, it exists. It contradicts the evolutionary idea that one occurrence arose from another and created mutations and offshoots over hundreds of millions of years. If the blob were our "archetypical thing," the first cell from which everything arose, we would not be what we are: humans.

Image 1. Web span of the *Caerostris darwini* (bark spider)

Image 2. *Segestria florentina* (green-fanged tube web spider)

Image 3. *Attacus atlas* (Atlas moth)

Image 4. *Oncorhynchus tshawytscha* (Chinook salmon)

Image 5. *Eretmochelys imbricata* (hawksbill sea turtle)

Image 6. *Amphiprioninae* (clownfish or anemonefish)

Image 7. *Balaenoptera musculus* (blue whale)

Image 8. *Tetraodontidae* (a young species of puffer fish)

Image 9. *Chamaeleonidae* (chameleon)

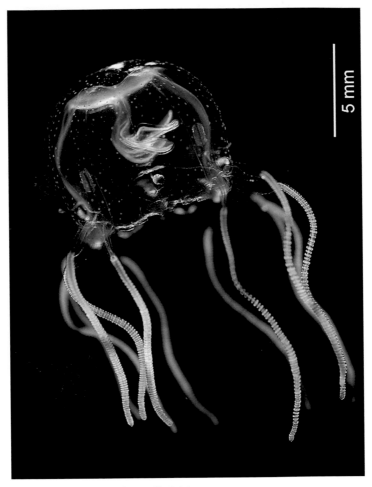

Image 10. *Cubozoa* (box jellyfish)

Image 11. *Coleoidea* (octopus)

Image 12. *Dionaea muscipula* (Venus flytrap)

Image 13. *Physarum polycephalum* (a species of slime mold)

Image 14. Traces of artificial structures on Mars

CHAPTER 2

SCIENCE!—
SCIENCE?

Without the invention of the printing press, we would know nothing of the theory of evolution. In German-speaking Europe, it is common knowledge that Johannes Gutenberg (1397–1468) was the inventor of the printing press. But in China printing with movable type was known over 1,000 years earlier. During the Han period, from around 202 BC to 220 AD, printed papers existed. But in Europe, it was Johannes Gutenberg from Mainz discovered the art of printing pages with movable letters. Nobody knows whether he used an old Chinese print product as a template or whether he independently invented casting letters in lead. Gutenberg worked on the so-called *Vocabularius ex quo*, a Latin-German dictionary that served as an aid to understanding the Bible. Then, between 1452 and 1454, he had the Bible printed in his own workshop in Mainz, a total of 150 copies on paper and 30 on parchment: the now world-famous *Gutenberg Bible*.

At that point, many more people than ever before were able to view the scriptures themselves. Discussions about God and creation reached larger circles and, inevitably, the question arose: How did man come into existence? Were Adam and Eve really primordial humans created by God? Thanks to the printing press, scholars from different countries learned what other gentlemen—at that time there were no learned women—thought about the origin of life. The Bible said that God created plants first and then animal life. The Greek philosopher and astronomer Anaximander of Miletus (610–547 BC) had the same thoughts. He traced the origin of humans back to earlier living beings. More than 1,000 years later, the Paris clergyman Denis Diderot (1713–1784), who also wrote novels, formulated the idea that humans could also descend from animals. Today, the majority of the scientific community is convinced that the theory of evolution has been definitively proven and that all questions about the various ways of life can be easily explained. Another group of scientists, which is becoming more important every day, takes exactly the opposite position: the origin of life and of species is in fundamental contradiction to evolutionary theory. Which is correct now?

The Briton Charles Darwin (see Figure 2.1) is considered as the inventor of the theory of evolution. But in his work *On the Origin of Species*, he named several scholars who, like himself, had come up with similar ideas.[1] In 1809, for example, the work *Philosophie Zoologique* by French botanist Jean Baptiste de Lamarck appeared in Paris and prompted the first dispute among experts on evolution. Lamarck advocated for the idea that all organisms would pass on the properties they had acquired during life. He named the giraffe's long neck as one of many examples.

According to Lamarck, the ancestor of the giraffe would have had to stretch their neck in a dry environment in order to reach the highest-growing leaves of tall trees. And that's why over generations, the giraffe developed a long neck. It was only decades later that various philosophers and botanists realized how dangerous "Lamarckism" could become for human society. If traits are inherited, then either smarter—or more stupid—people are bound to be born. A dictator could refer to his ancestors and claim that his present knowledge contains their knowledge. He could call himself superior to the others (blue blood) and that everyone should therefore respect him.

Figure 2.1: Charles Darwin

Finally, the German physician Friedrich Leopold August Weismann (1834–1914) attacked Lamarckism in several lectures. He demonstrated that the idea of inherited properties via cells could not be correct. For example, he asked, Why don't the soldiers in an anthill change? Of course, Weismann knew certain facts about heredity, but these did not arise from some mutation in the cells, but from inheritance via the germ line. Only mutations in the germ line can cause changes in the body, not the cells, even if they may have changed due to the environment.

Weismann had only initiated the dispute over Lamarckism. The real scientific debates began after his death. When French Baron Pierre de Coubertin (1863–1937) resurrected the Olympic Games in 1894, he was imbued with the spirit of "neo-Lamarckism." His idea was to make the French fit through sport. Just like the English of the time, he believed that fitness was inherited, and accordingly each generation would be more productive than the previous one. These ideas were also represented by the National Socialists of Germany. The results are known. So what is there about this evolutionary theory that is so controversial, and how did Charles Darwin come up with it in the first place?

Charles Darwin was born on February 12, 1809 in Shrewsbury, England. After school, which he completed in one of his mother's religious communities and in a boarding school, he studied theology and philosophy. As a 22-year-old, he was allowed to take part in a trip on the survey ship *HMS Beagle*. The *Beagle* belonged to the British Navy and the voyage was to go to South America, Patagonia, Tierra del Fuego, and Chile. Charles, who in his spare time had always been interested in birds and sea shells, was overjoyed. He brought a lot of writing materials and paper on board. The trip lasted 5 years. The *HMS Beagle* returned to England on October 2, 1836.

On the Galapagos Islands, Darwin took note of slight differences in turtles and birds. Each clearly belonged to the same species as the others, but they looked slightly different depending on the island and they also were performing different tasks. He observed finches with thick beaks and those with narrow beaks (see Figure 2.2). Some were able to break open nuts, others had to be content with seeds. Darwin soon realized why. The finches with narrow beaks were incapable of cracking nuts because they had grown up in a different environment—on a different island—than those with thick beaks. The environment had changed the species. Darwin noticed similar changes in a wide variety of animals and plants. So the thought grew in him that the stronger creature would prevail, according to the principle of the "survival of the fittest."

Figure 2.2: Various finches

Darwin also observed that the animals of a population always produced more offspring than were actually necessary for the conservation of their species. In addition, the animals of a genus could develop different skills depending on what their environment had to offer. Darwin soon realized that the animals that adapted best clearly had an advantage over the less adapted. And he was of the opinion that the different skills acquired by one species would be passed on to the next generations. The theory he eventually put forward was actually not fundamentally mind-boggling in his day. Darwin personally knew several scholars who had similar thoughts. Even so, he initially did not discuss his ideas in public. For a full 20 years, he discussed his views only among colleagues and in particular within the Linnean Society of London, of which he was a member. The Linnean Society is the oldest biology society in the world, and a certain Alfred Russel Wallace (1823–1913) was also a member. Wallace was an insect collector, anthropologist, and an excellent draftsman. So it happened that Darwin and Wallace first exchanged and presented their ideas in the Linnean Society: theories that, although developed independently of one another, had many things in common. Thanks to the art of printing, Darwin's theory of evolution spread around the world, sparking storms of enthusiasm on the one hand and hateful ridicule on the other. Where did the emotionally charged controversy come from?

Until Darwin's publication, the Holy Scriptures had been accepted as valid, and there—in the first book of Moses in the Bible—one could read how God created Earth, then the plants, the animals, and finally humans. Biblical teaching was undisputedly sacred, dictated by God personally. For believers, and this was true for

Christians of all groups, but also for Jews and Muslims, the heavenly Creator had created the world with everything that lived on it. Doubting this was considered heresy and one would have ended up burned at the stake 100 years before Darwin. But the minds of many intellectuals had been working hard for a long time. Darwin's time was the time of Enlightenment. His theory made possible the liberation from the constraints of belief. The power of the churches angered quite a few people—they wanted to know about things and not just believe. Two hundred years before Darwin, the British statesman and philosopher Francis Bacon (1561–1626) had already published several writings in which he called for a separation of knowledge and belief. Bacon was convinced that humans can only understand nature if they analyze the connections for cause and effect. Therefore experiments would have to be carried out and the connections between the events uncovered. The principle "knowledge is power" is often attributed to Francis Bacon.[2]

Darwin had also studied theology and philosophy, and so he was familiar with the publications of German philosopher Immanuel Kant (1724–1804; see Figure 2.3). He was considered an enlightener and demanded real proof of the existence of God. Kant had demanded courage from scholars and instructed them to use their own mind and to stand by their thoughts. Kant was just as familiar with the theory of the formation of our planets from a primordial nebula as he was with the fact that the planets revolved in orbits around the sun and Earth revolved around its own axis. He even thought inhabited planets outside of our solar system were normal. Long before Albert Einstein, Kant distinguished between space and time and lectured that the nature of objects "as they are in themselves" is unknowable to us but that knowledge of appearances, of what they look like, is nevertheless possible.[3] He considered the evidence of the existence of God put forward by theologians to be nonsense:

> *So when one brings the concept of God into the natural sciences and their context in order to make purpose in nature explicable, and afterwards in turn needs this purpose to prove that there is a God: then there is no inner essence in either of the two sciences.*[4]

IMMANUEL KANT
From a painting

Figure 2.3: Immanuel Kant

In his main work, *Critique of Pure Reason,* Kant called for the distinction to be made between sensations and reality, the subjective and the objective.

> *Everything that, apart from the good way of life, a person still believes to be able to do in order to please God is mere religious delusion and a pseudo-service (cultus spurius) of God.*[5]

Charles Darwin was also familiar with the teachings of Arthur Schopenhauer (1788–1860; see Figure 2.4) who considered himself a student of Immanuel Kant.

Figure 2.4: Arthur Schopenhauer

Schopenhauer, who held a doctorate in philosophy, took the view that the world was based on "an irrational principle."[6] Schopenhauer knew Johann Wolfgang von Goethe personally, and in Weimar they both had philosophical discussions. It is not known who influenced whom. Schopenhauer advocated for the idea that the will is the driving force of man. Every action always follows a will. He considered "the character of man" to be innate and unchangeable, against which nothing could be done, and only because of "character" could a person want certain things. Schopenhauer drew a clear distinction between understanding and reason. The mind can judge what is happening. It can determine the cause of a noise or the force with which a spear has to be thrown in order to hit its target. Rationality, on the other hand, investigates the reasons behind it:

> *Any stupid boy can crush a beetle, but all the professors in the world cannot make one.*[7]

Darwin's theory of evolution burst into this world of thought. The ground was prepared. The humanities scholars, all who counted themselves as such, rejected the biblical story of creation. It is true that people went to church, prayed with their families, kept the Ten Commandments and the holidays, but deep inside they were torn by doubt. What kind of God created plants and animals, and had qualities such as "omnipotent" and "all-good," but was unable to remedy the injustices of this world? Would a good-natured God create a nature out of plants and animals in which the stronger always ate the weaker, and often in a cruel way? The upper class of the time, who believed themselves to be better than the working people, were not happy with a religion that constantly threatened and put people under pressure with terms such as sin, corruption, and condemnation—a religion that condemned everything and considered everything having to do with sexuality unclean. For centuries, the church had not only increased its power, but had also abused it. Accordingly, it was not unusual for the intellectual class to doubt the biblical statements about the origin of the human race. There *had to be* other answers. Reason, not belief, should become the leitmotif of life. Therefore, people sought other ways of looking at the processes in nature. Darwin's doctrine of evolution and selection fit in perfectly with this. In addition, his evidence was visible to everyone and appeared rational. Darwin did not claim anything, he did not ask for belief, but he could present the different plants and varieties of animals of the same genus. What more could one want? Although Darwin was ridiculed in public and images were circulating that showed him as a monkey, his theory became increasingly popular.

The intellectual upper class applauded, first secretly, then outwardly. Any reasonable person could follow Darwin's line of thought, and anyone who condemned his theory was soon no longer considered "popular." Slowly, a new stratum of believers formed: those who believed in evolution. Anyone who rejected Darwin's theory was suddenly pitied with a condescending smile. They were said to still believe in the story of the Dear God of the Bible, who created everything. What ensued was to be expected: Darwin found brilliant advocates.

Ernst Heinrich Philipp August Haeckel (1834–1919; see Figure 2.5) was one of them. Haeckel, an outstanding physician and professor of anatomy in Jena, Germany, contributed significantly to the spread of Darwin's theory

Prof. Haeckel, Jena.
Taken with ZEISS Tessar

Figure 2.5: Ernst Heinrich Philipp August Haeckel

of evolution. Among other things he was the inventor of the term *ecology*—though today's Green party politicians don't seem to know this. As an anatomist, Haeckel knew of every cartilage and every bone in the human body, and based on his specialist knowledge, he formulated the natural history of creation.[8] Haeckel rejected the biblical act of creation. He postulated a unity between spirit and matter (a thought that is also advocated for by well-known physicists of our time). Behind the atom, the subatomic particles are recognized, and behind that is "the spirit in matter."[9] Haeckel supported the idea of teaching Darwin's theory as a subject in high school. This led to fierce controversies, some of which were held in public. As recently as 1878, the Social Democrat August Bebel (1840–1913) had campaigned in the Reichstag in Berlin to include Darwin in the school's curriculum. His opponent Otto von Bismarck (1815–1898) argued against it.[10]

According to Haeckel, the beginning of all life can be traced back to a common origin.[11] This too is a modern thought: today's science sees a common primordial cell behind the origin of life. But Haeckel developed more and more into a racist. He gave lectures on artificial breeding and human eugenics and became a pioneer of social Darwinism. Leading Nazi propagandists later referred back to Haeckel's work. In 1900, Haeckel was the chairman of a society that dealt with eugenics. There, the equality of people was classified as a mistake that would lead to a degeneration of the entire civilization. Haeckel wrote:

> *The killing of newborn crippled children, as practiced, for example, by the Spartans for the purpose of selecting the fittest, cannot reasonably fall under the concept of murder, as is still the case in our modern law books.*[12]

But fueling the debates that were being waged in all imaginable intellectual circles, clubs, and political circles at the time was Darwin's theory of evolution. It was the breeding ground, the salt of the new worldview, coupled with the loss of faith in the biblical creation story and the appeal to pure reason.

The *Communist Manifesto* (Figure 2.6) by Karl Marx (1818–1883) and Friedrich Engels (1820–1895) burst onto the scene during this time of rebellion. The idea of communism was born In the *Communist Manifesto*, the authors referred to Darwin: everything can be explained in a natural way, everything is reasonable. There is no need for creation. God is dead. Two main schools of thought emerged ever more clearly in society: that of the believers and that of the unbelievers. The believers clung to God, they believed in creation and that everything was permeated by a divine spirit. The unbelievers did not

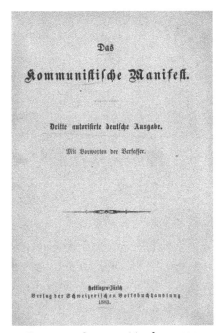

Figure 2.6: *Communist Manifesto*, 1883

believe in any of this. You didn't need a god to understand nature and life. And anyone who wanted to have a say and was still plagued by doubts about what was right and what was wrong read Nietzsche. He had finally done away with the Christian God.

Friedrich Nietzsche (1844–1900; Figure 2.7), a pastor's son with an excellent education, decided to condemn his Christian faith (Figure 2.7). Of course, Nietzsche admired Charles Darwin's theory of evolution. He was also a supporter of Schopenhauer, and Schopenhauer one of Kant. Nietzsche despised the believers. In his eyes, they were ignorant. Nietzsche had studied philosophy in Bonn and Leipzig. In 1869, when he was just 25 years old, he received a professorship for Greek language and literature at the University of Basel, Switzerland.

Figure 2.7: Friedrich Nietzsche

But 9 years later, he had to give it up for health reasons. Physically, Nietzsche was small and sickly; he suffered from severe headaches. Perhaps this was one reason why he glorified the so-called superhuman. In 1876, he published a treatise with the title *Untimely Considerations* and stated in it that the political and moral values were subject to the current zeitgeist.[13] Nietzsche was in fact correct from today's perspective.

Nietzsche was a friend of composer Richard Wagner. Nietzsche and Wagner spent time together on Lake Lucerne in Switzerland. But the friendship faltered later. Nietzsche described Wagner's art as sick. He told his former good friend that he was "groveling before the cross."[14]

Nietzsche's thinking was increasingly shaped by Darwin's theory of evolution. He pleaded for the man of power, a stronger species of men; he hated the weak. In his works *The Will to Power* and *Thus Spoke Zarathustra*, he describes the stages of human development: from dependency from which only the strong can free themselves to becoming free and independent thinkers. For Nietzsche, believers of every religion were grovelers or bootlickers, unable to rise. He testifies to this in his book *The Antichrist*, a merciless reckoning with his own religion. After all, as I mentioned earlier, Nietzsche had grown up as the son of a pastor and had attended excellent Christian schools. Nietzsche's choice of words and power of language are incomparable. A few quotes from his book *The Antichrist* may prove this:

> *No one is free to become a Christian. One is not "converted" to Christianity—one must first be sick enough for it. (Chapter 51)*
>
> *Faith means not wanting to know what is true. (Chapter 52)*

*This eternal accusation against Christianity I shall
write upon all walls, wherever walls are to be found—I
have letters that even the blind will be able to see—I call
Christianity the one great curse, the one great intrinsic
depravity, the one great instinct of revenge, for which no
means are venomous enough, or secret, subterranean and
small enough—I call it the one immortal blemish upon the
human race. (Chapter 62)*

*I condemn Christianity; I bring against the Christian church
the most terrible of all the accusations that an accuser has
ever had in his mouth. It is, to me, the greatest of all imaginable corruptions; it seeks to work the ultimate corruption, the
worst possible corruption. (Chapter 62)[15]*

Nietzsche celebrated his anger against God—especially the Christian one. According to Nietzsche, the
thought of an "inherited sin"—original sin—could only
arise in sick brains. And by the corruption of the church,
he meant the indulgence trade and the mendacious busi-
ness dealings with plots of land. Countless Christians had
bequeathed their possessions to the church only so that
their earthly obligations would be redeemed and they
would be accepted into heaven—according to Nietzsche,
this was nothing but lies and deceit.

Despite his attitude toward the strong man and the
man of power, Nietzsche completely rejected German
National Socialism. But Nietzsche's world would never
have come about without Darwin's theory of evolution.
Darwin was the first to create the scientific prerequisites
for making a god superfluous. Religion claimed the supernatural origin of the world. Now at last, there were natural
explanations for this. For the intellectuals, who wanted
nothing to do with the supernatural, this was liberation.

More and more groups demanded that humanity should build on physical evidence and not on unprovable belief. So, public—and published—opinion shifted towards science. Faith was a private matter; knowledge was power. Although many great minds suspected that the theory of evolution could never be the key to understanding the various species, they still defended it. They did this stubbornly out of the compulsion that the atheistic worldview brought with it. The French biochemist Ernest Kahane (1903–1996) said on November 17, 1964, in a lecture at the European Organization for Nuclear Research (CERN) in Geneva:

> *It is absurd and utterly nonsensical to believe that a living cell arises by itself. Still, I believe it, because I can't imagine it any other way.*

The Scottish anthropologist Sir Arthur B. Keith (1866–1955) put it even more clearly. Keith was a member of both the Royal College of Surgeons in London and the National Academy of Sciences in the US. In Siam (Southeast Asia, today mainly Thailand), he had studied monkeys and great apes for years, and he saw in the monkeys the origin of man. Though an ardent admirer of Darwin, he said in his inaugural address to the Rectorate of Aberdeen University:

> *The theory of evolution is unproven and unprovable. But we believe in it because the only alternative to it is the act of creation by a god. And that is unthinkable.*[16]

Many a biologist and anthropologist felt the same way. God was not allowed to exist; it was better to just believe in the impossible, just not in God. In the *New York Review* of

January 9, 1997, biology professor Dr. Richard Lewontin wrote:

> *We take the side of science despite the obvious absurdity of some of its constructs, because we cannot allow a divine foot in the door.*[17]

And Nobel Prize winner James Watson, one of the leading genetic researchers and co-discoverers of DNA, who, however, became a racist in old age, remarked:

> *The theory of evolution is a world-wide recognized theory, not because it can be proven, but because it is the only alternative to creation in which we do not want to believe.*[18]

At last, we should also quote atheist and philosopher Malcolm Muggeridge (1903–1990), a British journalist and public advocate of the theory of evolution:

> *I myself am convinced that the theory of evolution, especially the extent to which it is applied, will go down as one of the greatest jokes in the history books of the future. Posterity will be amazed how such a weak, dubious hypothesis could be accepted so incredibly gullibly.*[19]

In the meantime, Darwin's teaching has become the credo of anthropology. Any doubt about it is something of an academic suicide. At the beginning of the evolutionary dispute as well as today, two irreconcilable groups face each other. The "pro-Darwin" representatives, insulted by the other side as godless materialists, and the "counter-Darwin" representatives, seen by Darwinists as naive people who still believe in a dear God. In the US, the land of Christian religions, strong groups of so-called creationists formed. These are those who believe in creation. They reject Darwin outright. What is supposed to be wrong with the theory of evolution?

In the first chapter of this book, I presented the abilities of some animals that are difficult to fit under the mantle of evolution:

The power station in the belly of the electric eel

The gender reassignment of animals from male to female and back again

The snail with two penises

The tracking system of the bats

Cloned ant soldiers next to normal ants

The various different snake venoms

Water spiders that transport an air bubble under water

The "exploding" cnidocyte of the sea wasps

The brain of the octopus, which is distributed over the whole body

The trapping leaves of the Venus flytrap, which close up within a millisecond

The blob—a thinking and planning lifeform without a brain

All the questions that I have raised are also familiar to the most astute biologists and zoologists. They know the holes in Darwin's theory, they know about the absurdities. So, explanations have to be found for the impossible. And if no solution will fit, the theory will be adapted. It is stretched and broadened, always under the guise of science. Today's Darwin is no longer the original Darwin. Here are two examples:

Two species of the gastric-brooding frog (*Rheobatrachus*) lived in Australia. The southern gastric-brooding frog was first discovered in 1972 in Southeast Queensland, north of the city of Brisbane, and the northern gastric-brooding frog was discovered in 1984 in a mountain range of the Clarke Range in the north-east of the same state.[20] These frogs did the impossible: they bred their young in their stomach. How is that supposed to work? After insemination, the female swallowed the fertilized eggs and the stomach stopped doing digesting. It suddenly changed its function. The gastric juices dissolved and the organ mutated into a uterus. The frog could not eat any food during the breeding season. Up to 25 young frogs grew in her stomach. When they were ready to "hatch," they left their mother's body through the mouth, one after the other.

This kind of breeding in an animal's stomach is completely impossible in a slow, evolutionary process. The mouth is designed to eat and the stomach to digest; there is no other way. And yet the fact remains that the frogs hatched in the stomach. How do evolution theorists solve such a puzzle?

A similar problem arises with the development of the eye. Where did the first light-sensitive cell come from, from which eyes later developed? In their excellent work *Evolution, Ein Kritisches Lehrbuch* (Evolution: A Critical Textbook) biologists Reinhard Junker and Niko Winkler note that "The eye prototype cannot be explained by selection, because selection can only advance evolution if the eye functions at least to a small extent."[21]

By what miracle did evolution produce an eye? After all, evolution or nature cannot know in advance which organ will one day be important. Evolution is not a thinking mind, planning for the future. So evolution theorists bridge their dichotomy by assuming that evolution is probing. Sounds good and is just as unscientific as the opinion that there is a god behind it. A probing, that is, trying evolution, would again presuppose some kind of mind/spirit or someone supernatural that/who can do the probing. It's like this: Let's see how it turns out.

One simply adapts. Claus Wedekind, Professor of Ecology and Evolution at the University of Lausanne, Switzerland, commented: "Evolution is verifiable; it does not necessarily have to last millions of years. Evolution can sometimes happen very quickly."[22] In the 1930s, Nobel Prize winner Thomas Hunt Morgan (1866–1945) developed the first additions to Darwin's theory. Morgan worked as a professor of experimental zoology at the California Institute of Technology, in the US. He thought Darwin's theories were fundamentally correct, but still too speculative. So he looked for evidence and began studying the Drosophila fly in depth. He discovered the positions of the various genes in the fly's chromosomes.[23] Morgan is now considered to be one of the founding fathers of the new science of genetics. A synthetic theory of evolution emerged, according to which developments always take place in continuous, small steps and that inheritable influences on the genome are not possible. The changes of the species were defined in the order of the genes.

This theory also had to be expanded, and the critical theory of evolution and the systems theory of evolution emerged. In the critical theory of evolution, it is not the environment that brings about the evolutionary changes, it is the genetic composition or structure achieved in each case. Even this theory cannot explain the genesis of something fundamentally new in the genetic sequence. The systems theory of evolution also did not help to clarify the elementary questions. According to it, the genes placed higher up in the genome should be responsible for changing subordinate genes.

Evolution theorists had been familiar with the general terms macroevolution and microevolution since 1927. Macroevolution is the development of new forms, such as a completely different organ in a living being. Microevolution is the change within a species—for example, when different dog species emerge from one dog. The safe harbor hypothesis was also developed; this proposes that the development of a lifeform takes longer in a safe environment (development stage) than in a dangerous environment. For example, predators' craving for food would force a larva to develop more quickly than in a peaceful environment. How the larva is supposed to know that it is being eaten and has to pass on its information to the next generation remains inexplicable. Social evolution is no less ingenious. It is correct that people also adapt to their social environment, but the next generation and the one after that can grow up in completely different social environments, for example because of wars or devastating natural disasters. Social behavior is not passed on in an evolutionary sense. According to the original teaching of Darwin, populations that do not adapt are unlikely to be able to survive in the long run, but they will nevertheless continue to exist.

New ideas—new helper explanations—new theories. This led to evolutionary developmental biology—Evo-Devo. Researchers examine how individual development is controlled in evolution. Sounds confusing? It is. Because Evo-Devo is the basic idea for a whole series of additional thoughts on how *what* exactly could be explained. Evo-Devo is supposed to prove that the game of evolution contains far more organized processes and less accidental development than assumed. According to the opinions espoused by Evo-Devo research, embryonic development can give rise to spontaneous mutations and thus something new, whereby Darwin's selection (choice) only begins to have an effect afterward and selects the most suitable individual beings.

Darwin's idea focused on survival of the fittest, but Evo-Devo wants the arrival of the fittest to be clarified. How do the fittest come about? To understand this, new thoughts are built into the evolutionary system, all of which are very scientifically formulated and yet remain only auxiliary models. It is assumed that evolution prevents, or stops, changes that are too radical. But how should the blissful evolution know in advance what will turn out to be "too radical" a change in the future?

Evolution makes use of the idea of a "genetic toolbox" added by theorists; this is comparable to a switch that can turn lines of development on or off. It all sounds highly scientific, but it doesn't answer the original questions. How should "evolution" know which tool to take from the "box?" And if all remains coincidental, "the box" is of little help.

Evo-Devo research also deals with the so-called gene knockout, the switching on and off of certain genes. It has allowed growing functional eyes in living things that naturally have no eyes at all. For example, those genes of a mouse that are responsible for the growth of an eye were used in a fruit fly. And an eye developed on the fruit fly. But how the first light-sensitive cell came about remains a magical mystery. Evo-Devo also deals with gene sequences that are alike and still occur in various lifeforms; in other words, genes that in some way move from species to species. Whole sets of genes can be found several times, although the animals are not related to one another. How is that possible?

It is undisputed that the biologists are doing a great job. How could you explain *whatever* needs to be explained? What makes sense? Which gene combination causes the decisive changes? Do cells communicate with each other? If so, what signals are used? It remains to be seen how such signals came about—if they exist at all. How does the environment, stress, or laziness influence a way of life? How is the genetic information passed on to the next generation? And so on! The subject seems inexhaustible.

Research results raise new questions, and in the end, there is only one requirement: scientifically unambiguous proof that can be replicated at any time.

In 2006, the book *The God Delusion* by Richard Dawkins was published. From 1995 to 2008, he was professor of Evolutionary Biology at the University of Oxford, England. He had the following printed on the cover of his original edition:

> *I am against religion because it teaches us to be satisfied with not understanding the world.*

Dawkins is the most radical proponent of evolution and arguably the most furious prophet of atheism. He writes with the eloquence and clarity of Friedrich Nietzsche. For Dawkins, God is pretty much the stupidest thing that humans have come up with, and evolution is the most sensible thing. The following quotations prove it:

> *God is a vindictive bloodthirsty ethnic cleanser, a misogynistic, homophobic racist, an infanticidal, genocidal, phillicidal, pestilential, megalomaniacal, sadomasochistic, capriciously malevolent bully.*
>
> *It is absolutely safe to say that if you meet somebody who claims not to believe in evolution, that person is ignorant, stupid or insane. Atheism is almost always evidence of a healthy, independent mind.*[24]

The former Darwinists who, for good reasons, did not want to acknowledge anything divine or allow anything spiritual, have meanwhile become believers—believers in a theory full of contradictions, stubborn to all objections, and especially blind. They refuse to even acknowledge scientifically proven facts. Anything spiritual is unthinkable to this group of scientists and can never be a subject of science. A term like *love* would not fit into their thinking box.

What does this term *empirical,* which natural science demands, really mean? (Without empirical research, natural science wouldn't exist.) What is meant by *empirical research* is any kind of research in which the conclusions are made up of verifiable evidence. *Empirical* refers to data that can be replicated or observed at any time in experiments or in nature. Empirical research is progressing step by step. First, the purpose of the investigation must be determined: What do you want to find out? Then a hypothesis is construed and experiments are carried out. Then the data is analyzed and the final result is announced. How can the theory of evolution be proven empirically?

In its analyses, science always takes into account the rule of Occam's razor. What is that? William of Occam (1288–1347) was a theologian, philosopher and also a monk. He wrote several papers on logic, epistemology and even metaphysics.[25] Included in his works is a book with the title *The Principle of Simplicity*, although today this principle is often referred to as Occam's razor. (Immanuel Kant also referred to this.) Occam's razor demands that a simple explanation is always preferable to a complicated one. Example: In stormy weather, a fallen tree lies in the street. The simplest explanation is that the wind knocked the tree over. But it could also have been a meteor, an elephant, or a ghost. But meteors rarely fall, there are no elephants in Europe, and there are no ghosts at all. The explanation that the wind knocked over the tree is the most obvious. So, the more complicated options fall under Occam's razor. Basically, the rule is sensible, and it leads to the most obvious solution, but it can also lead to the most stupid solution. Come again? The next sensible solution is always subject to the current zeitgeist. Anyone who observes a moving light in the night sky concludes, according to Occam's razor, that it is a meteor. But it could also—in the case of a more modern zeitgeist—be a man-made space station. If a mountaineer feels sick and weak, he is naturally given medicine that is good for the stomach. But it's possible that the man's blood could also have been contaminated by radioactive radiation because he—knowing nothing and not suspecting anything— could have repeatedly crossed a vein of uranium ore as he hiked. But radioactivity can only be measured in our time. The next danger in the use of Occam's razor is coming up with a simple answer might discourage or block future research. You already have "the solution"; why should there be further research? And it is not uncommon that the "simplest solution" is stubbornly enforced, even when it becomes clear that it was incorrect. The concept of the "simplest answer" is also abused by political or religious groups. Often not everything is as simple as one thinks.

To provide empirical evidence for the doctrine of evolution means starting at the beginning. After all, everything has its origin somewhere, and at the beginning of evolution stands life. Here one can start empirically. Where did the first lifeform come from? How did it arise? Such questions can be empirically investigated using today's scientific methods. What is a cell made of? Which components does it contain? How did these components come about, and how did they come together? How did the components— the "toolbox"—grow into something living? Something that multiplied by itself, became more and more gigantic, and passed information from cell to cell? Was the new science of genetic engineering, which successfully identified the contents of a cell and, so to speak, looked inside the cell, not able to provide the empirical evidence for the origin of life?

But in order to be able to distinguish all the inner workings of a cell, you first have to know whether the cell contains anything at all. In fact, the first pioneering discovery was made in 1929 by biochemist Phoebus Aaron Theodore Levene (1869–1940). Levene was raised in Saint Petersburg, Russia, and emigrated to the United States in 1883. There, he first worked as a physician and later as a biochemist at the Rockefeller Institute in New York City. With the tools available at the time—the so-called super microscope—he studied the inner workings of the cell and discovered DNA (deoxyribonucleic acid), a macromolecule made up of the four basic bases—adenine, guanine, thymine and cytosine. DNA occurs in all cells and is the carrier of genetic information.

Initially, the majority of scientists rejected this newly discovered DNA. It took another two decades before the resistance was eventually overcome when the development of the electron microscope made Levene's discovery visible to everyone.

Twenty years later, biochemist Erwin Chargaff (1905–2002) from Columbia University in New York was able to prove that the DNA of every living being contained the same amounts of adenine and thymine as well as cytosine and guanine. Two of these basic building blocks always appear in pairs. Chargaff was the first scientist to explain the molecular appearance of the inner workings of a cell. He wrote several doctrines: 1) DNA always consists of the four basic nucleotides. 2) DNA samples from different tissues of an individual are the same. 3) The composition of the DNA is independent of age, nutritional status, or the habitat of a species. But where did this DNA come from? (At this point I have to repeat what I discussed in an earlier book.[26])

The DNA is a giant molecule (a molecule consists of atoms that are affixed to each other). On the basis of Chargaff's work, in 1953 Francis Crick (1916–2004) and James Watson (born 1928) succeeded in proving that the four basic bases of DNA stick together like a twisted rope ladder, comparable to a spiral staircase that has wrapped around itself—hence the term double helix. You can imagine the whole thing as a zipper that is twisted in a spiral. Each zipper has spikes. They correspond to the four basic building blocks adenine, guanine, thymine, and cytosine.

The DNA strand (rope ladder) lies in the core of every cell—without exception. Although all cells contain this DNA, it is not arranged in the same order in every form of life. There are microscopic differences. The DNA of a person with kidney damage is lined up differently "on the zipper" than a person with only four toes. This applies not only to the individual, but also to entire ethnic groups. Although it always remains human DNA—we are all brothers and sisters—we are still slightly different from one another. It is undisputed that a person born in China has different facial features than a Native American. To deny this on the grounds of alleged racism is completely unscientific. The basic characteristics of people in different parts of the world are known. We all remain human, no matter what skin color we have, but we are not the same. "Gender" does not undo this scientific fact either. Today, the science of genetics has mastered the technology of changing the genetic code—the order of the four basic bases in the DNA molecule—and can literally create other forms of life.

So how did the first cell come about? And the inner workings of this one cell? And, extrapolating from the answers to these two questions, did different forms of life evolve from the first cell over hundreds of millions of years, as Darwin's theory of evolution suggests? Can Darwin's teachings be empirically proven using the new and very exact science of genetics?

First of all, one thing should be made clear: the DNA in the cell is *no lifeform*. It's just molecules stuck together. These do not multiply like a cell, for example. They are made up of thousands of atoms. The different chemicals cannot simply "attach" themselves at any point of the DNA. The giant molecules form shapes with indentations and spikes. They do not consist of the same geometric images such as, for example, a square or triangle. Any number of molecules *cannot* "connect" with others; this has been empirically proven. A lock without the corresponding key to unlock it is worthless. Both have to fit. In DNA, only certain basic bases fit into the sequence. Others cannot dock at all. Nobel Laureate Jacques Monod (1910–1976) postulated in his book *Zufall und Notwendigkeit* (*Chance and Necessity*) that the molecules would bind to each other by themselves over the course of millions of years.[27] He said that the first cell was created by chance plus unalterable rules of nature:

> *Man finally knows that he is alone in the indifferent vastness of the universe from which he emerged by sheer chance.*

And Manfred Eigen (1927–2019) as well as Nobel Prize winners like Jacques Monod defended the opinion that ultimately it is universal laws of nature that control every chance.[28] Don't these statements make everything plain and indisputable? Who dares to contradict the influential great minds? They announced that the first cell had come about by chance but that this wasn't by chance at all because it was based on an unalterable regularity (Monod: *Chance and Necessity*). And God is dead anyway. The law behind chance is random.

Other luminaries wanted to know more about it. *How* exactly does *what* exactly happen in a cell? The great number of experts in research departments of large universities and chemical companies were equipped with X-ray crystallographic microscopes, which can actually be used to determine the composition of molecules. With the help of the so-called diffractive patterns, researchers were able to verify the position of every single atom in the molecule. Even the composition of tiny proteins became visible. Then, the phenomenal technology of nuclear magnetic resonance (NMR) was developed, which allowed an even deeper insight into the inner workings of the molecules. This also applied to the composition of the DNA. Now everything became empirical: replicable and controllable at any time. And the microbiological discoveries annoyed Darwin's disciples. It quickly became apparent that every cell is capable of producing thousands of different molecules and installing them in the right place in a hundredth of a second.

Evolution theorists since Darwin had argued that the eye emerged by way of a slow process of development. At some point, a light-sensitive pigment cell had formed, then a retina, around it a pupil, so that the incidence of light can be regulated, and also a diaphragm that focuses the light. Muscles developed that could move the eye quickly. The eye would send colored light signals to the brain and the brain would organize these signals. Darwin, too, accepted the complexity of the process and helped himself by postulating that, after all, there would also be animals with eyes that were simpler than those of humans. This proves the predecessors—i.e., the history of development—of the eye. And true to the principle of Occam's razor, this declaration was sufficient—at least until today's high-performance microscopes were used. Now, unexpected complexity has been revealed behind the eye. As soon as a glimmer of light hits the retina, a whole series of molecules react within microseconds; and they change in a flash; that is, the arrangement of their atoms changes. A so-called ion channel opens, which sorts sodium and calcium ions and only lets a certain amount of them through. Precisely defined proteins are switched on and off until a weak current flows from the optic nerve through a cell membrane into the brain and its evaluation results in images. Prof. Dr. Michael Behe (whom I will talk about later) commented on the events in the eye:

All those steps and structures that Darwin believed were so simple have to do with astonishingly complex biochemical processes that cannot be whitewashed with rhetoric.

And:

Chance belongs to the category of metaphysical speculation; scientific explanations, however, rely on causes.[29]

Chandra Wickramasinghe (see Figure 2.8), holder of several doctorates and today director of the Buckingham Center for Astrobiology at the University of Buckingham, England, analyzed the so-called chemical evolution at a conference in Sindelfingen, Germany, in front of an audience of 3,000: the inner workings of cells. He demonstrated convincingly that Darwin's theory is no longer tenable.[30]

Fred Hoyle (1915–2001), a professor of astronomy and astrophysics at the University of Cambridge, England, for many years, notes that:

In pre-Copernic times, the Earth was mistakenly believed to be the geometric and physical center of the universe. Nowadays, the respectable science sees the Earth as the center of the universe. An almost unbelievable repetition of the earlier error . . . Evolution can only be explained if the genetic material for the origin of life came from outside our system.[31]

Figure 2.8: Chandra Wickramasinghe

Fred Hoyle and his colleague Chandra Wickramasinghe published a paper in which they too came to the following conclusion:

> *The "home work" stage never existed on Earth. Life had already developed to a high level of information before the Earth was even formed. When we received life, all basic biological questions were already resolved.*[32]

This continues in the specialist literature. Even once die-hard proponents of Darwin's theory, such as David Horn, former professor of anthropology at Colorado State University in the US, now admit that chemical evolution could not have taken place on Earth.[33] And let's not forget Bruno Vollmert (1920–2002), professor of molecular chemistry at the University of Karlsruhe, Germany, who at the end of his many years of research, equipped with the most modern microscopes in the world, confessed:

> *The fact that DNA, and thus life, could not arise by itself . . . but if, on the other hand, life is clearly there, so that pieces of the DNA chain can also be analyzed and reconstructed in*

*the laboratory at any time, means that it must either have
always been there—which is not the case— or it owes its existence to intelligent, purposeful planning ... Darwin's doctrine of the origin of the species and of life in general through
mutation and selection is a great error.[34]*

One would think that the empirical results of the
biochemists and geneticists would cause a rethink in the
scientific community. But the opposite is the case. Ideological camps have formed. Clever scholars, all of them
with integrity and honesty, who often even worked and
researched at the same schools and universities, ridicule
one another. As soon as Thomas Nagel published some
fundamental contradictions of the theory of evolution,
while a professor of Philosophy at the University of California, Berkeley, he was ridiculed by his colleagues.[35] The
same happened to Michael Behe, Biochemistry professor
at Lehigh University in Bethlehem, Pennsylvania, US.[36] In
Haeckel's, Nietzsche's, Marx's and Engels's times, it was
still about getting rid of God, building on facts and isolat-
ing the metaphysical as stupid belief. Today, it is the other
way around: the Darwinists have mutated into believers—
stubbornly defending their theory. They are not interested in new findings. Occam's razor from the fourteenth
century is all they need to be satisfied. Today's keyword is
ideology. The biochemists and geneticists who dismantle
the components of the cell on their super microscopes
are dismissed as creationists, that is, as those who still
believe in a god. Most of these researchers have nothing to do with religion or belief in God. The other side,
represented by those who stubbornly cling to Darwin, is
accused of materialism. Opinions are bogged down and
resemble two erratic opponents. One group doesn't
read the other's publications because they already know

everything much better. And each group looks with pity at the other, which "just doesn't want to understand." There is a mentality of "why should I bother with this nonsense." This attitude has completely turned around since Darwin's times. Back then, people looked condescendingly on the naive "who still believe in God"; today, they look with pity at the stupid ones who still have not understood Darwin.

In the autumn of 1996, Michael Behe's bestseller *Darwin's Black Box* was published. Behe, who was born in 1952, began his scientific career at Drexel University in Philadelphia, Pennsylvania, with a Bachelor of Science. He followed this with four years of research in biochemistry at the University of Pennsylvania, followed by a further four years of research on the structure of DNA at the National Institute of Health in Bethesda, Maryland. He was then appointed Assistant Professor of Chemistry at Queens College, New York City, and in 1985, he was appointed Professor of Biochemistry at Lehigh. Behe is, in the truest sense of the word, a profound expert on the cell. Hardly anyone has carried out as many experiments and investigations into the inner workings of the cell than Behe has. In his 500-page book, he takes the unequivocal opinion: the cell is the product of some kind of planning:

> There is no publication in the scientific literature—in journals, specialist magazines, or books—describing how the molecular evolution of any real, complex biochemical system took place. There are claims that such evolution took place, but not a single one is supported by relevant experiments ... It can really be said that the claim about Darwinian molecular evolution is baseless chatter ... Why do many biochemists still consider Darwinism credible? The answer must be

sought in the fact that the students in their biology studies were taught that Darwinism corresponds to facts.[37]

Professor Behe was "made a complete mockery of" by his colleagues. His thorough research was disqualified as pseudoscience. He himself was accused of unscientific procedures and of course accused of being a creationist.[38] It's no surprise that the eloquent Richard Dawkins pounded his colleague Behe wildly and unobjectively in his book *The God Delusion.*[39] It became obvious that students *must* not be allowed to learn about Behe's ideas. The Darwinists feared that their faith might be lost. (By the way: Michael Behe is an ordinary Catholic and has absolutely nothing to do with creationism.) I will use an example to present his work.

What would happen if any container of any liquid becomes leaky? The liquid runs out. Depending on the type of liquid, it might run out faster or slower. But on a related note, what happens when a person injures himself and begins to bleed? He only bleeds for a short time, then the blood thickens and a plug blocks the blood from flowing out. The wound heals. Behe explains:

> *Blood coagulation is a highly complex, intricately networked system involving a large number of interacting protein components. If one of these components is missing, the system will not work.*[40]

Behe points out that a wound must close quickly, otherwise the person will bleed to death. This cessation of the bleeding must only be concentrated on the wound itself—a small stab or a larger cut. If the blood were to clot in the whole body or in other places, this could lead to the collapse of the body due to a heart attack. On pages

128–142 of his book, Behe analyzes the incredibly confusing system of different molecules that cause blood to clot only at the area of the wound, while most of the blood continues to circulate in the body. And every time a wound develops, entire armies of antibodies form in the body and prevent bacteria or even viruses from penetrating. There are thousands of these antibodies, all of which remain inactive until a foreign bacterium suddenly enters the bloodstream, including firmly adhering antibodies, messenger proteins, and mobile antibodies. In addition, it all works within a timeframe. None of the components can intervene at the wrong moment, for example *before* the other. Behe notes that a cell couldn't gradually develop such a system, as according to Darwin's theory. Everything had to be working together at the same time, like parts of an engine. Behe claims:

> *No one at Harvard University, no one at the National Institutes of Health, no member of the National Academy of Sciences, no Nobel Prize winner—no one at all can give a detailed account of how the cilium, visual process, blood clotting or any other complex biochemical process could have developed according to Darwin's rules . . . Darwinian theory is unable to explain the molecular basis of life.[41]*

Of course, Behe knows all of the literature, including the textbooks of his specialist colleagues who have dealt with similar topics. He is familiar with the reviews of his work, and he has answered critics without exception—even in highly scientific specialist journals. But that didn't stop the attacks against him. Why not?

Professor Behe committed the sin of all sins: he takes the position of Intelligent Design (ID). What is that? Intelligent Design is assumed to be intelligent planning. Someone or something—a spirit of the universe or aliens (?)—are behind this planning. The very thought of it is considered unscientific in the traditional scientific world. Positive publications about it are mortal sin. During a thought process lasting for centuries, starting with Francis Bacon, Immanuel Kant, and Arthur Schopenhauer, and then from Ernst Haeckel, Karl Marx, and Vladimir Ilyich Lenin, to Friedrich Nietzsche, this hated "God" finally had been gotten rid of—and now new types started popping up, chatting about Intelligent Design. Terrible! Burn them at the stake! God was dead—there couldn't be anything spiritual out there, including aliens that had created DNA. Biologist Reinhard Junker put it this way:

> There is a fundamental problem with the design approach in the fact that the actions of a creator and his identity are in principle not scientifically comprehensible and his approach cannot be scientifically described.[42]

That sounds factual—but it's not true. The scientific evidence has long been *available*, but it is being ignored. There are skulls of aliens on our good old Earth, empirically provable—I reported about it in an earlier book.[43] And there are people living among us who have an alien implant. There are also publications on this with empirical evidence (photos).[44] Whoever really wants to know about it today *will have seen* the UFO films even recognized by the US Navy. (I reported about it in my book *Botschaften (= Messages) from 2118*, starting on page 29.)[45] But the dogmatic guardians of the Grail from yesterday do not want to take note of it. Their previous world must remain in order. How does Professor Behe justify his change of heart towards Intelligent Design?[46]

The result of these systematic efforts to study the cell—the research of life at the molecular level—is that design is unequivocally attested. The result is so clear and so significant that the corresponding efforts must be counted among the greatest achievements in the history of science . . . Life was planned by an intelligent being.

The aforementioned Richard Dawkins, one of the grail keepers of the previous teaching, who disqualifies all those who do not believe in Darwin as stupid, categorically forbids the idea of Intelligent Design (ID). Its access to science is prohibited. ID does not belong in biology or physics class: "We must never be tempted to see these works as scientific treatises."[47] Nothing against other opinions—but Mr. Dawkins lacks the latest information.

Two hundred and fifty years ago, clergyman William Paley (1743–1805) had already asked the same questions about creation as today's researchers do on the subject of Intelligent Design.[48] Paley said that one should imagine a person who has never seen a clock or heard of a clock in their life—and now, suddenly, finds such a thing in a meadow. The amazed person would bring home what they had found. It would be admired and cautiously taken apart. This would reveal small cogs, a spring, a winding mechanism, two rotating hands, and so on. The society of that time would come to the conclusion that there must be beings living somewhere in the world who had created this marvel. Nobody would get the absurd idea that the complicated object had come about by itself.

There is planning behind the clock and there is understanding behind the planning. But we constantly pretend that everything came about "naturally." We seek and find the stupidest excuses to rule out planning behind

creation. When we see a crane, it is said that "nature" provided it with stilted legs *because* it cannot swim.

"Nature" gave it a long beak *because* it had to look
for food in the water. There are thousands of similar
explanations for evolution. Modern evolutionary theory
is teeming with nonsensical reasons, all of which were
invented in order to be able to fit the theory, so as not to
encourage further questions. It is constantly assumed that
omniscient "nature" wanted this or that *because* a lifeform
needed it. The omniscient "evolution" gave the crocodile
a skin armored with scales on its dorsal surface *because* it
needed one in its environment. As if with this perverse
logic all other animals in the same environment did not
have the same needs. A land animal moves into the water
and mutates into a whale—this one now has nostrils on
the top of the head *because* it is more practical in the water.
The same whale gives birth to its young under water and
has to maneuver them to the surface to breathe, otherwise they would drown. Why didn't this outrageous
"nature" also ensure that the animals can breathe under
water, *because* after all, the most important thing, survival,
was at stake? Bats developed an incredible tracking system
because they need it in the dark. But other animals also live
in the dark—and without the bats' tracking system. Chameleons change their appearance in a flash *because* this
is necessary for survival. How about some color-changing
mice? They also have to survive in the same environment.

The different snake venoms developed because of the
different animals various snakes eat. Oh holy stupidity! A
spider weaves gigantic threads of up to 25 meters (more
than 80 feet) in length, and the blissful "nature" ensures
that these threads are made of the hardest biomaterial—
because otherwise the threads will tear. Wow. You really

have to think hard about that one. The sea wasp's stinging cells literally explode when they come into contact with an enemy because this is how the animal keeps the predators away. As if this didn't affect other sea creatures as well. Certain species change their gender from male to female and back again because this process is needed for the development of the young. I'm sorry, what? Have you ever thought about that a little longer? We confuse cause and effect. We don't want to acknowledge that there is something outside of our sensory world that we have not yet discovered. That's basically pure chauvinism. Originally, the word chauvinist meant an exaggerated nationalist. I am the best! I beat anyone! Today's Darwinists, who have already done away with God, do not tolerate Intelligent Design either. There had to be something hidden behind this attitude that is more profound than that. In depth psychology, this is called *repression*. Behe comments:

> *Lots of people, including many distinguished and respected scientists, just don't want to accept there exists anything beyond nature. They don't want a supernatural being to influence nature.*[49]

And this type of chauvinist dictates to others what scientific thinking is and what should never be part of it. The claim that intelligent design is unscientific because the designer behind it cannot be grasped is nonsense and only exists in the minds of those who do not want to look at the modern empirical data on it. The zeitgeist is no longer the same as it was twenty years ago. Today we know that there are trillions of planets out there. We *know* that billions of them are Earth-like. We *know* that extraterrestrial lifeforms exist, have visited us in the past and keep watching us. We *know* that huge distances in the universe can be bridged. We *know* that in addition to the innumerable other forms of life in the universe, there are also human-like forms. At least whoever wants to know, knows. With this contemporary knowledge, it is dishonest not to include Intelligent Design in science just because the designer(s) cannot be captured empirically. This is comparable to the example of the clock found in William Paley's story: because the clockwork is clearly artificial and had to be planned and manufactured, one must not infer an intelligent planner from it. Heaven help us!

But how on earth should you imagine Intelligent Design? Who are these designers? What could their motives be? How did they go about it? In my earlier books, I kept building bridges of thought in order to come closer to the answers. Now I would like to present a type of situation—without any claim to accuracy, but with this request: just think about it!

Imagine that somewhere out there in the vastness of space an Earth-like planet existed with beings like us. However, these strangers would be 100 years ahead of us in terms of their technology. They also managed to extend their lives considerably. Each of them can easily live 400 years. That's why I call them "the ancient ones." Just like us, they evolved, learned to catalog species relationships, and developed phenomenal technologies. There was only one point where they got stuck: How did life arrive on their planet? They developed high-performance microscopes, examined the inner workings of the cells and were faced with the same catalog of questions we humans are. Where did the program in the DNA come from? After decades of profound debates, the scientific elite finally agreed that the origin of life must be found in outer space. However, they did not share this knowledge with their citizens because, like on Earth, several major religions were present on their planet. And the idea of an intelligent, alien species was incompatible with religious righteousness.

It was clear to the intellectual elite on the planet of the ancient ones that since life had come from outer space, those unknown strangers might one day turn up again. Would they exterminate the population of the planet of the ancient ones? Exploit them? Enslave them? How should one prepare for such an eventuality?

The ancient ones developed giant telescopes and listened for signals from space. They sent satellites to their neighboring planets and looked for the origins of life, the amino acids. In fact, they found such building blocks of life, but they also discovered bacteria, and—what was even more astonishing—they found traces of former building structures on some planets. This proved that unknown beings must have worked there at some point, because the ancient ones themselves had never traveled to those planets, so the structures had to have come from someone out there.

One day, the unbelievable happened: the telescopes on the planet of the ancient ones registered two small objects that were flying toward their planet. The astronomers reflexively assumed that they were splinters from meteors. But the objects performed artificial maneuvers; they were intelligently controlled. The surprise was huge and the excitement within the astronomical community led to riots.

What was to be done? Should they shoot down the objects with rockets? The ancient ones decided to send a satellite on the trajectory of the alien objects to get closeups. At the same time, they tried to radio those alien objects on the firmament with laser light and radio waves on all frequencies. One afternoon, the screens in the control center began to flicker, and a pale, narrow face appeared with huge eyes, a thin nose, and a small mouth. The stranger seemed to be smiling, although his mouth was clearly toothless. Then the ancient ones heard a voice and they hardly dared to breathe. The alien spoke in their language. He said, "Greetings, brothers," and repeated the words several times.

No scientist on the planet of the ancient ones had ever expected such communication. They were talking over each other until it finally dawned on them to install a camera and microphone and answer on the same wavelength. One of the ancient ones was elected to speak, and he smiled at the camera, nodded a few times like he had been trained for it and said obediently: "Be greeted, brothers." The word brothers had been agreed on because the strangers had used it. Next, the ancient ones wished to know why "the brothers" knew their language. The toothless stranger replied: "We learned your language from your electronic media. You made it easy for us to communicate."

A brief dialogue ensued, from which the inhabitants of the planet of the ancient ones learned nothing. An overwhelming majority in the committee of the ancient ones was of the opinion that their population was not prepared for such an encounter. Planet-wide riots could break out. The strangers out there communicated little, but invited a group of eighteen ancient ones to visit their spaceship. So, a delegation was put together. It consisted of three astronomers and three astrophysicists, two biologists, three high-ranking theologians of the great religions, two philosophers, two diplomats, two secretaries for the protocols and a writer. A small airport on the edge of a desert was chosen unanimously as the launch site. The population shouldn't notice anything.

After only 40 minutes of flight, the small spaceship docked onto a gigantic circular object: the aliens' mother spaceship. It turned slowly on its axis, on the dark side of the moon. The first sight came as a real shock to the ancient ones. The mother spaceship had to have a diameter of at least 4 kilometers (2.5 miles)—previously unimaginable for the ancient ones. The brothers had provided rooms with the right atmospheric conditions, and weeks of intense discussions began. Evening after evening, the brothers brought their eighteen visitors back to their planet, and they had no objection when the ancient ones replaced members of their delegation.

The ancient ones learned from the brothers that the universe is infinite and eternal. Whole galaxies would collapse and merge in space and time, new Big Bangs would explode and other galaxies would be born. They compared the universe to an endless sphere—it has neither a beginning nor an end. The question of the origin cannot be answered. They do not know how or even when the universe began. They believe that behind it is something they call the "spirit of creation."

The ancient ones' scientists had always been of the opinion that some aliens would have to have completely different body structures than they did themselves. After all, evolution develops different forms of life on other planets. Why did the brothers look like the ancient ones?

The brothers proved to be patient and very accommodating teachers. "That was correct; our bodies are similar," they replied, "but each species would migrate its own throughout the universe." Now, however, the distances between the stars are so huge that migration with spaceships is not feasible. That is why sectors of a galaxy are infected with their own building blocks of life. Their spaceship, they said, had released trillions of DNA molecules into space. DNA molecules are resistant to the vacuum and the cold of space.

Up to this point, the ancient ones had understood everything. But what then? How should the DNA molecules know which planet were suitable for them? The brothers lectured objectively and demonstrated their words in color 3D pictures.

The ancient ones learned that the number of suns and planets in the universe is infinite. Among them there are also planets of the same type as theirs. Most of the distributed molecules are taken in by the gravity of a sun and evaporate. Others get caught in the gravity of completely unsuitable worlds, where it was too hot or too cold, or where they encountered gas giants or micro-moons. All of this is insignificant. Only a small fraction of the original molecules land on a suitable planet. And that's where evolution begins.

"You have proven that on your world as we have on our own," they said. "So," one of the delegates of the ancient ones asked, "are there completely different-looking forms of life out there compared to us?"

"Of course; it depends on each specific world. There are species with two heads and six tentacles, but they could not survive on your planet. They might breathe ammonia. Others move with suction cups and a recoil principle. They could not exist on this planet either. Your environment is too hot. Every species that conducts space travel at some point looks for destinations where it can live and expand. Or do you want to land on a gigantic planet? The gigantic forces of gravity would crush you to a pulp and the atmosphere would be highly toxic. So you look for related worlds. You want to survive. The principle applies to the entire universe."

"We *are*," said one of the philosophers with a deep sigh, "and you *are*. What is life in the universe for?"

The lecturer of the brothers stretched out his two arms and moved them in a circle. Holographic solar systems emerged from nowhere. The eyes of the ancient ones could barely follow. The whole conference room seemed to be made up of suns and moving planets. The lecturer's voice, though slightly higher than that of the ancient ones, quivered:

"The spirit of creation wants expansion. The entire universe should communicate. The intelligence should penetrate all dimensions and times. The purpose of all of us is to colonize matter with the vibration of intelligence. This occurs through the power of curiosity. It is the driving force behind the expansion. That is why the Spirit of Creation provided the first life. The most diverse lifeforms would arise on an evolutionary path; including those with curiosity. Every curious lifeform will ask questions; it wants to know more and more and more. What's out there? How did we come into existence? Why did the all of this happen? What is the origin? Every life should gather its own experiences, exchange information and expand. Endlessly—until that entire universe is filled with intelligence. We assume that only then will we understand the Spirit of Creation. This will be the day of knowledge. But now the number of planets that could support intelligent life is infinite. It would be impossible for a single species to fly a spacecraft to all of them. Therefore, a very large number of cultures is required who also conduct space travel. In your language, brothers, there is the phrase snowball effect. The intelligence in the entire universe is supposed to multiply based on the snowball effect. We will support you."

There was silence. While the holographic solar systems in the conference room slowly faded, one of the astronomers whispered to the philosopher: "Weird—in our ancient traditions, we can read that the gods created us in their own image. Now I understand." The astronomer raised his head and turned to the alien lecturer: "And how did the different animal species on our planet come about?"

The answer came promptly: "Within the species, this happens through the evolutionary processes known to you, outside the species through the information of other beings. We are not the only species that spread DNA in the universe. In addition, new species are being created in order to test their competence of adaptation and their chances of survival on distant planets—like research in a laboratory. And you should learn that your world has never been a closed system. This is true throughout the universe. No planet within an ecosphere is a closed system. All planets are open to outside forces."

The profound information from the brothers shook the ideological picture of the ancient ones. It was only a short time before the leaders of the religions decided to unite.

The population should no longer be separated by a sense of self-righteousness but should unite by the common reverence for the Spirit of Creation. The curtain of immaturity was lifted in small spurts. And within a few years, the first spaceship will take off from the planet of the ancient ones. They are now part of the universal snowball effect.

Of course this is an invented story. Nothing else. Nothing else? Somewhere in the story there is a religious touch. I began to wonder how a large community like Buddhism actually thinks about evolution. So I asked the Japanese Zen master Ryofu Pussel for an answer.

Buddhism and Evolution

Ryofu Pussel

The historical founder of Buddhism is called "Buddha," but this is a name of honor given to him later: it means the enlightened one. Its original name was Siddhārtha Gautama. He was born 566 or 563 BC—April 8th is generally accepted as his birthday—as the son of a prince of the city of Kapilavastu (in today's Nepal), about 500 years before Jesus. According to legend, Māyā (the name of Buddha's birth mother) could not conceive children from her marriage to Prince Shuddhodana for 20 years. But one night, a sacred white elephant penetrated her right side. She got pregnant. Her husband was delighted. At the end of her pregnancy, she traveled to her parents' home to give birth, but gave birth on the way there. Her baby Siddhārtha was able to walk and speak right after he was born. His mother is said to have ascended to Tusita heaven. Buddha later married, but gave up his life as a prince when he was 29 to become an ascetic. After he became enlightened through his steady meditation practice at the age of 30, he was henceforth called Buddha—the enlightened one. In this context it is not unimportant that he visited his mother Māyā in Tusita heaven for 3 months after his enlightenment. The Tusita heaven is a divine world. In addition to other deities, Buddha Maitreya also resides in it. This is the Buddha of the future and world teacher—he is to return to Earth to teach us in 5,670,000,000 years after the historical Buddha.[1] Time passes by differently in Tusita heaven: a Tusita day and night correspond to 400 Earth years, a Tusita month to 12,000 Earth years, and a Tusita year to 144,000 Earth years; the life span of a Tusita being averages 576,000,000 Earth years.

What does Buddha teach about evolution? 2,500 years ago, he already proclaimed the theory of the "Big Bang," millennia before the same idea was taken up by Western astronomers. In Buddhism, "after a long time, there finally comes a time when this world ends . . . Then, after additional long periods of time, the time will come when this world will develop again."[2] And in the Anguttara-Nikāya text, Buddha introduces "Eon," a unit of measurement; an eon corresponds to the period of time it takes for a world system to come and go. Furthermore: Our universe consists of endless world systems that are harmoniously dispersed in space. There are both individual planets in it but also those that form a group by the thousands; there are also "super-galactic" groupings of galaxies. Within this seemingly endless universe there are other inhabited worlds and systems in which beings go through life and death cycles. Buddha went on to explain that just as humans and other living beings go through the cycle of birth and death, this also affects these world systems (galaxies). Each of these galactic evolution cycles is one such eon. An eon is the time it would take to remove an 11 kilometer high mountain made of pure granite, using a cloth of the finest yarn that only touches the mountain once every 100 years.[3] Prof. Gethin from the University of Bristol compiled a table of Buddhist cosmology based on Buddha's statements. In the two lowest worlds there are animals and humans. Above that are six worlds of the lower gods (evolutionary cycles up to 128,000 "divine" years), then three worlds of the higher gods (evolutionary cycle up to 1 Eon), the worlds of the pure form and the formless form with evolutionary cycles of up to 84,000 Eon.⁴ Buddha explains that these enormous periods of time are perceived as short during the journey itself. One wonders whether this describes time shift effects when traveling at extremely high speeds, as Albert Einstein postulated. For example, 50 human years correspond to a "divine" day, and 9 million human years are even 500 "divine" years.⁵

For Darwin, Homo sapiens developed by evolution; in Buddhism, humans are descended from "God." In the second part of the Aggañña Sutta, Buddha explains it as follows: Here we develop, but the "divine" spark is in us, and it distinguishes us from inanimate matter. Darwin talks about our bodily and physical development, Buddhism also includes the spiritual and Social part, connected to psychological development. "Everything is in continuous change," says one of the main teachings of Buddhism. Darwinism sees everything "in mutual dependence." In Buddhism, evolution takes place over huge periods of time. But in both Buddhism and Darwinism, humans and animals are part of the common system of life in this world. Buddhism goes one step further: We are basically a product of evolution. But even more: the "divine" spark, the "Buddha-nature" (Sanskrit: buddhadhātu) is in us. This is found in all of the teachings of the historical Buddha. For example in the Mahāparinirvāna Sutta, which contains the overall summary of his teachings; there, in chapter 12 one learns: All beings have Buddha-natures. This is the real self . . . It's strong . . . and this "divine being" is indestructible . . . Buddha-nature is inseparable from our self . . . And the power of the gods participates in evolution. The path to connection with the "divine" is contained in each one of us. This is the teaching of Buddhism.

Sources:

1. Japanese-English Buddhist Dictionary (Tokyo: Daito Shuppansha, 1999).

2. The Book of the Long Texts of the Buddhist Canon, Section "The Teaching Lecture about Knowledge of the Ancient Times," in Agganna Sutta in the work Digha Nikāya: Section DN 27 DN iii 80; Translator: Dr. R. Otto Franke. See Sutta Central, https://suttacentral.net/dn27/de/franke, Accessed: April 27, 2020.

3. Peter Harvey, An Introduction to Buddhism. Teachings, History and Practices (Cambridge: University of Cambridge Press, 1990).

4. Rupert Gethin, "Cosmology and Meditation: from the Aggañña Sutta to the Mahāyāna," History of Religions 36, no.3 (Chicago: University of Chicago Press, 1997).

5. Harvey, An Introduction to Buddhism.

I did not study Buddhism, nor did I belong to any Buddhist community. I read Buddha's reflections on evolution in the article by Ryofu Pussel for the first time. The kinship of our thoughts was very perplexing to me.

On the basis of my previous list of the strange abilities of animals (in Chapter 1), some things about evolution can definitely be questioned. And what about us humans? Are we just the product of evolution, or is there more to it? That is the subject of the next chapter.

CHAPTER 3

HUSHED UP AND
SUPPRESSED

Luis Navia, who held the chair for philosophy at the New York Institute of Technology for 28 years, is well versed in all the evidence for and against evolution. His opinion is as follows:

> *There are scientists who speak of evolution as a fact and are immediately ready to condemn the "mysticism" and "pseudo-science" of others. Perhaps they are not aware that scientific hypotheses can only become facts when we know for certain that all possible relevant information has been analyzed.[1]*

Biochemist Oliver Mühlemann from the University of Bern, Switzerland, sees this completely differently:

> *Where do we come from? How did life come about? In the last 200 years, the natural sciences in particular have provided plausible answers to these questions, which are so central to our self-image and our worldview. Everyone, first and foremost Charles Darwin with his simple and elegant theory of evolution, which has been confirmed thousands of times. It*

explains conclusively how today's diversity of living beings developed from common, identical primordial cells. "²

Two brilliant and witty scientists—two opposing views. Is science becoming a dogma? Which of the two do you have to "believe" now?

Anthropology is the specialty that concerns "what we know about man." The word *anthropology* is made up of the Greek words *anthropos* (man) and *logos + logie* (knowledge). The name comes from German philosopher Magnus Hundt (1449–1519) and is 400 years older than Darwin. Originally, the term *anthropology* was mainly applied to the descent of humans. Today there are subdivisions such as philosophical anthropology, historical anthropology, theological anthropology, psychological anthropology, cultural anthropology, and even cybernetic anthropology. The generic term *anthropology* always stands for a group of human sciences. The development of humans in special areas is examined. But in general, anthropology is one of the natural sciences. In fact, it should deliver exact results, and its evidence should stand up to empirical criteria.

And that's exactly what it lacks. In Darwin's anthropology, an image is glued together on the basis of skull and skeletal bones to make the case for the ancestry of the human being, but this evidence is all open to challenge. As in every profession, in addition to thoroughly honest anthropologists, there are also those who cheat. And unfortunately many—too many!—who simply refuse to take note of all of the arguments against Darwin's doctrine. The most brilliant and detailed work in which the dishonesty in anthropology is exposed in minute detail is that of authors Michael Cremo and Richard Thompson:

Forbidden Archeology.[3] Shortly after the publication of the
1,000-page book in the United States, the television station
NBC produced a documentary to draw the public's attention to the possible errors in the theory of evolution. But
the American anthropologists complained vehemently to
the US regulatory agency, the Federal Communications
Commission, and the film was only allowed to be shown in
a censored form. It is hard to believe: in the freest country
in the world, where freedom of speech is constitutionally
guaranteed, there is a censorship authority. It prevented
evidence *against* Darwin's teachings from being made
widely available. One is involuntarily reminded of the
dogmatic behavior of religions. The other side *must not*
be right.

Authors Cremo and Thompson, the latter has a PhD
in Mathematics, examined skeletons and stones, analyzed
skulls in museums and those that had been deliberately
destroyed on site. They followed the life stories of the
respective researchers and examined curious objects—
that should not actually exist—for their authenticity. The
result was sobering:

> *If all the available evidence is viewed with an open mind, we
> have to conclude that no evolutionary picture of the human
> origin arises from it . . . it is almost impossible to say any-
> thing about the origin of man . . . it is obvious, that pre-
> conceived notions about human evolution have played an
> important role in suppressing reports of unusual stone tool
> production. This is still the case today.*[3]

In November 1938, Wilbur G. Burroughs, then a
geologist at Berea College in Berea, Kentucky, wrote an
article in the college magazine. In it he claimed, together
with some colleagues, to have found traces of beings on a

petrified sandy beach in the Rockcastle district, Kentucky, which must have been human-like.

> *The footprints are imprinted on the horizontal surface of a hard, massive gray sandstone. There are three pairs of footprints showing left and right footprints . . . each footprint has five toes and a pronounced bulge.*[5]

The site and the footprints were certified by several experts. Geologist C. W. Gilmore confirmed that the imprints lie in a rock layer that definitely belongs to the Upper Carboniferous, which was utterly impossible. The age of the Upper Carboniferous was around 320 million years ago, and back then, there could be no human-like beings who also walked on two legs. An ethnologist noted that the footprints were probably carved in the ground by members of a Native American tribe. So the prints were examined microscopically. The images showed grains of sand between the toes and on the heels that were more tightly packed than the grains next to the feet. This demonstrated the pressure of the body on the heels and toes. In addition, the recordings under the microscope did not show any traces of any artificial processing. Nevertheless, the entire specialist group of anthropologists declared that the prints must *be* forgeries, because no bipeds existed in the Carboniferous Age. *Science News Letter* wrote in its June 1939 issue.

> *We admit that we don't know exactly how the prints were created, but we do know that it couldn't have been one particular creature, and that is man in the Carboniferous Age.*[6]

And in the January 1940 issue of the magazine *Scientific American*, Albert Ingalls wrote:

If you ask a scientist about the people in the Carbon age, it's like asking the historian about diesel engines in ancient Sumer. Science knows that these prints are not from a person in the Carboniferous, unless 2 plus 2 is 7 or the Sumerians had planes and radios.[7]

Thanks to Darwin—the opinions have been formed.

In 1850, a church was built on the hill of Colle del Vento near Savona, Italy. Construction workers came across a human skeleton at a depth of 3 meters (almost 10 feet). Its bones lay in a natural environment, embedded in a characteristic sedimentary rock made of clay and limestone. In the same layer, bones from a rhinoceros and marine clams also appeared. The layers clearly belonged to the Middle Pliocene—over 4 million years in the past. The skeleton found belonged to a person of small stature, and it lay in geologically completely untouched and intact layers. Later excavations uncovered fragments of different animals in the same layer.

The find was made 180 years ago. Rumors arose that it was a burial place of the Sisters of Mercy of Savona. In 1871, a conference for prehistoric anthropology was held in Bologna. The priest, D. Perrando, gave a lecture on the skeleton and proved that the place of discovery could have never been a burial place of the Sisters of Mercy. Over the years, more and more parts of the skeleton simply disappeared. In today's textbooks on anthropology, the find is hardly mentioned, and if it does appear sporadically, it is commented on negatively and labeled as manipulation. Which—according to Cremo and Thompson—must clearly be seen as publishing of false statements.

In the same series of deliberate misleading belongs
the case of Hueyatlaco, which I reported on in an earlier book.[8] The town of Hueyatlaco is 120 kilometers (75
miles) southeast of Mexico City. In the fall of 1960, workers came across several oddly shaped stones. The experts
consulted clearly identified the stones as tools. Now the
objects were dated using four different methods: 1) the
uranium series method, 2) fissure trace dating, 3) the
tephra hydration method and 4) the mineral weathering
method. All of the independent dates showed an age of
250,000 years. But 250,000 years ago there couldn't be any
man-made or human-used tools, at least not on the American continent. So, the stone tools were not accepted by
the anthropologists. Cremo and Thompson commented
on the case as follows:

> *The problem lies much deeper than Hueyatlaco. It concerns
> the manipulation of scientific thinking by way of suppressing
> puzzling data . . . Hueyatlaco was rejected by archaeologists
> because it contradicts the theory. This is a circular argu-
> ment. Homo sapiens developed around 30,000 to 50,000
> years ago in Eurasia. It is therefore impossible to have tools
> that are 250,000 years old that could be traced back to Homo
> sapiens, since Homo sapiens only emerged around 30,000
> years ago. That way of thinking makes for complacent scien-
> tists, but lousy science.*[9]

The stubbornness of too many anthropologists cries
out for a new science. The old one is obviously unwilling
to examine the latest findings without prejudice. If an
anthropologist deals with extraterrestrials or even with
Intelligent Design, he does it with the superior smile of
the know-it-all: It's all nonsense anyway! At the end of
their book, Cremo and Thompson show a table with over

120 finds, neatly listed according to the age of the objects, the location of the find, the category (bones, stone tools, footprints, minerals, etc.) and the reference number. Clear and easily accessible. But the professionals who should actually tackle it shrug their shoulders at best.

The "impossible" dates of, for example, 320 million years (footprints in Rockcastle, Kentucky) or 30 million years (skeletons near Savona, Italy) strikingly remind *me* of the equally "impossible" dates of the ages in Buddhism. And even *before* Buddhism, the Jain religion proclaimed similar, "impossible" ages. Their scriptures tell of people who worked 8,400,000 years ago and of others who reappeared every 100,000 years on average.[10] And only as a cross-reference, other than religious literature, I would like to remind the reader of the Sumerian list of kings. It is now in the British Museum in London, and according to it, the ten ancient kings ruled a total of 456,000 years before the great flood. The Maya in Central America had a god named Bolon Yokte. According to a tablet in the temple XIV of Palenque, Mexico, this god first appeared on July 29, 931,449 BC. On the third tablet in the Temple of the Inscriptions in Palenque, a date is carved into the stone in connection with the boy king Pakal, which is 1,274,654 years in the past.[11]

We do not even take note of this and countless other dates (for example in Egypt) because it is assumed they cannot be true anyway. Have we all fallen victim to the evolutionary worldview? *Can't* we think differently because we were taught this way and not otherwise? For the generation of my grandfathers, fathers, and my own, it is perfectly reasonable that ape species existed before humans. What if facts contradict this? What if there were people millions of years ago? People living next to apes?

It goes without saying that we assume that our great-great-ancestors, who chiseled the "impossible" dates in stone, printed them on clay tablets, scribbled on parchment or papyrus, were wrong, were wrongly informed or grossly exaggerated to make their rulers appear larger than they really were. But all of these are nothing but insinuations—assumptions—that have been made by us. Thousands of years ago, however, writing was a sacred art that only a few mastered. At that time, neither papyrus nor stone were entrusted with lies. The ancient ones knew very well what they were saying. We are the know-it-alls.

In his book *Die Evolutions-Lüge* (The Evolution Lie), Hans-Joachim Zillmer comments on several scientifically documented cases of dinosaur bones that do not fit anywhere in the anthropological model.[12] The fossils are far too young. According to general doctrine, all types of dinosaurs died out due to a meteorite impact around 66 million years ago.

But the dinosaur bones in question were dated around 24,000 years ago. Still, we are told it's not possible.

In the river bed of the Paluxy River near Glen Rose in Texas, fossilized prints of dinosaurs were discovered next to those of human feet. Not possible. No human can ever have met a dinosaur. For this reason, the prints of humans and dinosaurs in the same geological layer are dismissed as fakes. They are ignored in the scientific literature. Forgery is the magic word with which everything inappropriate is readily rejected. Just like the shoe print found in June 1968 near Antelope Springs, Utah. I reported about it 44 years ago.[13] But nothing has changed in the anthropological textbooks since then. Fake. But it is not a fake. Here's the story:

On June 3, 1968, William Meister stayed with Francis Shape, his wife, and their two daughters in the Antelope

Springs area, 43 miles from Delta in the state of Utah. Meister, armed with a hammer, was a dedicated fossil hunter. This was his hobby. On that day, he didn't find anything himself, but the girls did. They found something on a rock that looked like an elliptical curve. Meister carefully hammered away the stone all around it, until suddenly a layer peeled off, "like a page in an open book."[14] When the accomplished collector held the strange layer in his hand, he began to doubt his five senses. Two prints of human shoes were clearly visible in the stone. There were neither heels nor toes nor arches of the feet, but instead clear edges of pointed shoes: 32.5 centimeters long, 11.25 centimeters wide and 7.5 centimeters (12.7 × 4.4 × 3 inches) at the heels.

The heel of the left foot had crushed a trilobite, the remains of which were petrified together with the shoe prints (Figure 3.1). Trilobites are so-called primordial

Figure 3.1: Petrified footsteps

crabs that roamed Earth about 500 million years ago. William Meister took his curious find to Melvin A. Cook, a professor at the University of Utah, who, in turn, recommended that he consult a geologist. He was speechless and said that the find must be a fake because 500 million years ago, no one on Earth walked around in shoes.

Since then, the shoeprint together with the trilobite has been examined several times for authenticity. The trilobite is definitely in the print, and the rock layer is 500 million years old. Everything is real. Which doesn't change anything about the scientific doctrine. Even the opinion of evolutionary biologist Robert Martin from the Field Museum of Natural History in Chicago, Illinois, brought no change. Martin is convinced that humans and dinosaurs lived at the same time:

> *The primates that include humans originated about 90 million years ago. So the ancestors of gorillas, chimpanzees and humans lived side by side with the dinosaurs and did not evolve after their death.*[15]

Several molecular genetic studies by Martin and his team confirmed his statements. (As a side note: There are millennia-old finds around the world in the form of pictorial representations depicting humans and dinosaurs side by side, carved by Stone Age humans who allegedly never saw a dinosaur.[16]

A shoe print discovered in a coal seam in Fisher Canyon, in Pershing County, Nevada, also belongs to the same type. The imprint of the shoe sole is so clear that even the traces of a strong thread can be seen. The imprint is 15 million years old and doesn't fit into the evolutionary line any more than the shoe and footprints mentioned earlier.

In January 2020, over a thousand scientists signed an online petition critical of Darwin's teaching, including scientists from the US National Academy of Sciences and those from the famous universities of Yale, Princeton, Stanford, MIT, and University of California, Berkeley. The petition, entitled "A Scientific Dissent from Darwinism," expresses skepticism about the claims of random mutations and natural selections.[17] "The complexity of life has to be taken into account," it claims. To put it more clearly: the previous evidence for Darwin's theory should be carefully examined. "The only remedy against superstition is science." This statement comes from British historian and chess master Henry Thomas Buckle (1821–1862). Applied to the theory of evolution, the statement is no longer correct. Even the most recent finds from our present day are ignored there. Finds that were readily available for verification. The finds have something to do with the origin of intelligent humans. Because if it can be proven that people created tools 30,000 or 50,000 years ago, scribbled on stones and provided rock faces with ingenious drawings, then the previous sequence in evolution is incorrect. The lore that 10,000 years ago only primitive cave dwellers existed is plain fantasy. Here are a few examples:

I read a report in the German magazine *Focus* that electrified me.[18] I investigated the matter, found it confirmed, and wrote about it.[19] The reaction? A rather angry phone call from a professor of prehistoric history whom I personally knew who said that no importance should be attached to such finds. What was it all about?

The French astronomer Chantal Jègues-Wolkiewiez and friends visited the rock paintings in the Lascaux cave in the Département Dordogne, France. The pictures show horses, deer, bulls, handprints, and meaningless lines and points. Everything made in the colors available to Stone Age humans. The archaeologists saw nothing else behind this but the need of the hunters of the time to beautify their caves. But Madame Jègues-Wolkiewiez noticed completely different correlations.

In reality, the "meaningless points and lines" that archeology had always assumed to be around 5,000 BC showed entire constellations, first mentioned by the Babylonians and Chaldeans. But the cave paintings were at least 18,000 years old, 13,000 years older than the Babylonians. Madame Jègues-Wolkiewiez put together a map of the starry sky as it would have presented itself to the onlooker 18,000 years ago. Then, all points and lines were precisely measured and the results were compared with the starry sky from 18,000 years ago using a computer program. The matches were perfect. The astronomer Gérard Jasniewicz of the University of Montpellier, France, confirmed:

Several elements are beyond reproach. The orientation of the cave according to the solstice, the positioning of the Capricorn, Scorpio and Taurus in the hall correspond to the starry sky at that time.[20]

And how did anthropology respond? It called it pure speculation.

In September 1991, rock carvings were found in the Henri Cosquer Cave off Cap Morgiou in the Mediterranean Sea (southeast of Marseille, France). But the only entrance to the cave and thus to the paintings is 37 meters below sea level. Scuba divers with special cameras photographed and mapped the images; color samples were brought to the surface and later dated. The paintings are dated between 19,000 and 27,000 years ago. The pictures showed bison, penguins, cats, antelopes, a seal, and inexplicable geometric symbols. Since our Stone Age ancestors knew no diving suits and definitely did not draw paintings under water, all of this must have happened when the Mediterranean level was at least 37 meters (~120 feet) lower than today. When was that? A question that needs to be answered by geologists. And how did the people of the Stone Age in the Mediterranean know penguins? Were they traveling as globally active Stone Age hikers? I also suspect that the dates have been set as way too recent.

The same question marks apply to the underwater ruins off the island of Malta. I had referred to the so-called cart ruts in *Odyssey of the Gods*.[21] These are rail-like furrows in the ground that run in parallel curves across the island of Malta and then sink into the depths of the Mediterranean. Now the divers Thorsten Morawietz and Ramon Zürcher—the latter is my research secretary—dived for these cart ruts. At depths of 10 to 40 meters (30 to 130 feet), they repeatedly came across blocks cut into rectangles, some with gradations.

In the area of the sunken city Belt fil-bahar, the monoliths become more and more powerful the deeper one dives. Here, huge cuboids lie like a row next to each other at a depth of 28–35 meters. Countless monoliths lie there on the ocean floor. Sometimes it can still be seen that they broke apart.[22] See Figures 3.2 and 3.3.

Ruins under water can also be found in the Atlantic, for example, at the coast of the city of Carnac in French Brittany, or in Lixus, Morocco, as well as in the distant Pacific with the basalt buildings of Nan Madol (Caroline Islands), near the southern tip of Yonaguni (Japan), or off the Indian coast near Mumbai (exact position: 22nd degree latitude, 14 minutes east, 68th degree longitude, 58 minutes north). A whole city is under water there. What do you know about it?

Figure 3.2: Cart ruts under water in Malta

Figure 3.3: Cart rut rock

Among the many battles described in ancient Indian texts, especially in the gigantic work *Mahabharata*, there were also those between the Yadu dynasty and a "demon" named Salva.[23] Salva asked the demigod Shiva for a "heavenly vehicle" that would be able to fly anywhere. Shiva agreed and commissioned his chief designer Maya (the same one who also builds vehicles in the epics and puranas) to create a flying city. It must have been a terrifying form of flight, and it could move at such a speed that it was almost impossible to see with your eyes. This vehicle was supposed to destroy the city of Dvaraka, the headquarters of the Yadu dynasty. In and under the city, strong antiaircraft weapons were installed. The battle lasted twenty-seven days. Thousands of chariots were destroyed, and thousands of elephants and tens of thousands of warriors died. While the city of Dvaraka broke up piece by piece, an airship of the god Krishna appeared in the firmament. It headed for Salva's celestial vehicle to stop it. But Salva fired a mighty projectile at Krishna, which shone incredibly brightly. Krishna, for his part, showered Salva's sky city with a veritable flood of projectiles that outshone the entire firmament like suns. Finally, Krishna tore the sky city of Salva apart so that it fell into the sea in the form of thousands of pieces. The town of Dvaraka lay also in ruins. Today, both the remains of that ancient city and the fragments of the heavenly battleship of Salva lie on the seabed. Underwater cameras were used on site, then magnetometers and underwater metal detectors. First, the cameras' lenses captured artificial blocks of stone changed by human hand, "which, because of their size, could not be transported at all."[24] Then walls appeared that stood in right angles to each other, streets and the outlines of former buildings. Everything must once have been part of "a very advanced civilization." Nail-like pieces of metal with a high proportion of silicon and magnesium were secured, and the Indian scientists never doubted that there must be many more metal pieces lying around on the Dvaraka seafloor. This was confirmed by measurements conducted with metal detectors. The scientific report on the finds, which are up to several hundred meters off the coast, concludes with the words:

> *The references in the Mahabharata to Dvaraka's city were neither exaggeration nor myth. It was reality in the truest sense of the word.*[25]

Indian geologists who participated in the investigations also found remains of walls under water that clearly showed traces of rock glazing. At some point there must have been a terrible temperature here.

The historical traditions are known and so is the exact geographic position of the archaeological sites. Here is an opportunity to demonstrate the truthfulness of old stories. In addition, one could prove the existence of an advanced technology thousands of years ago because the presumably rust-free remains of that spaceship are still untouched in the silt of the seafloor. Why don't Indian scholars continue? It's always about the money. All clever financiers have their advisors, and they in turn say that terrifying truths from the pre-historic past should be left there. Humanity is not ready for it. Is "science?" Why don't universities take action on their own initiative?

Probably because "science" doesn't know anything; neither about the structures underwater off Malta nor about those off the coast of India. Books by me or my colleagues are not read in those circles. And the scholars who do, remain silent. You cannot take sides. Your status does not allow this. Changes of a scientifically firmly anchored opinion can only take place at the snail's pace of a generational change. This is due to the system. Every scientific journal checks the submitted manuscripts according to certain criteria. A work is only published after several experts have given their okay. On the one hand, this procedure protects the relevant specialist community from a flood of irrelevant publications; on the other hand, it blocks a discussion of potentially explosive discoveries. With regard to Darwin's view that one way of life developed slowly from another, all counterarguments have faded for decades. They are not picked up at all.

In his book *Die Evolutions-Lüge* (The Evolution Lie), Hans-Joachim Zillmer points out that a slow adaptation of lifeforms from water to land is not possible. A fish— according to Zillmer—cannot survive out of the water for more than a few minutes:

"If generations of fish had tried to visit dry land, they would all have died in a few minutes." The reason for this lies in the "complex organs such as a fully developed lung" that cannot come into existence all of a sudden. "However, a slowly developing lung is not functional in every intermediate stage, which is required according to the theory of evolution." According to Dr. Hans-Joachim Zimmer, "There has never been a partially or half-developed lung and, of course, it cannot be found in the fossils."[26]

Evolution theorists assume that some primordial crab has slowly, over millions of years, adapted to the conditions on land. If so, Earth should actually be teeming with the corresponding fossils. But they don't exist. However, this argument about the lack of fossils also applies to the slow development from some type of ape to humans. Two or three bones don't prove anything. We should have an incredible number of fossils.

The doctrine assumes that jellyfish and polyps have developed into flatworms, annelids, leeches, and finally crabs. At some point, vertebrates emerged from this. Which would have been the crab's first mating partner? Intermediate stages that served no purpose do not work. Did the mutation happen to change an entire crab population? The amphibian, the first reptile, emerged from crab and bony fish. But the same amphibians are said to have gone back into the water to lay their eggs—so says the theory of evolution. The last amphibian—or the first reptile—of this line of development was what anthropologists called Seymouria. This creature is said to have been the missing link between amphibians and reptiles. But where did its first mating partners come from? Without those, the line would have died out again. It goes on and on: New species appear, new populations are given new names, even if the animals had no mating partners and their chromosome numbers must have been completely different. Let's not forget the countless forms of life that came and disappeared over millions of years—completely without our knowing.

In the Proceedings of the National Academy of Sciences of the United States of America (PNAS) one can read how, in our time alone, "millions of species" are threatened by extinction.[27]

More than 540 terrestrial vertebrate species are said to have been wiped out in the twentieth century. Each dying species causes "a domino effect that brings further extinctions with it."[28] It was no different millions of years ago. Species came and went; where were their traces? As far as the origin of man is concerned, anthropology is content with a few bone finds that are thousands of kilometers apart and that were acquired from different continents. So, every couple of years, some fossils appear that have been declared our newest ancestors in the media. Once you were satisfied with *Homo erectus* (upright), the next thing that was praised was *Homo habilis* (gifted person) and then *Homo sapiens*. New skulls, new offshoots, new names . . . and every ape offshoot should be proof of our Darwinian descent. If *Homo sapiens* once descended from the Neanderthals who at some point descended from the Australopithecines, this theory is again not true because new discoveries disturb the clear-cut lines. It is never about any cemeteries with mass bone finds, but always about individual pieces that are far apart. The whole thing is sold as secured knowledge. Even the infallible *Encyclopedia Britannica* reports that there cannot be the slightest doubt about the fact of evolution.

Is that really true? The doctrine of evolution justifies the change from ape-like creature to humans with some flimsy approaches. The nonsensical use of the word *because* followed by reasons for the next lineage are more diverse than the few that I have already cited. It is assumed that *because* the early humans lived in packs, they developed social behavior; a kind of responsibility toward the other and their own group.

This is said to finally have led to the emergence of intelligence—planning for the future. This is nonsense. Many animal species lived and live in packs. Gorillas and chimpanzees do this as well as lions and schools of herring. Their intelligence is limited. Or: The prehistoric man climbed down from the trees for climatic reasons because he could move faster on the ground than in the trees. And because the food supply on the ground was more diverse than in the branches. This is nothing but a dud of an idea. As if some monkey species discovered something that worked to their advantage, but none of the others in the same family did it. They still jump around in the trees today.

Equally absurd is the claim that humans have no fur *because* they have learned to clothe themselves with other fur. As if our great-great-ancestors didn't lose their body hair until they began to get dressed. Even more stupid: With less body hair, you wouldn't sweat as much. The sweat that escapes would evaporate more directly through the skin. And should that only apply to our line—that of *Homo sapiens sapiens?* Why do bears wear thick fur? And not just the polar bears, but also those in the warm zones of world like Arizona? And why didn't the gorillas lose their fur when they live in hot and humid areas? Or the cat-like animals that live all over the world? I even read somewhere that we had lost our fur because skin without hair cools faster in water. Therefore, it is argued, hippos and elephants, for example, have no fur. Hippos couldn't survive with fur. It would cause heatstroke. And the elephant could not splash cool water on its body. All just excuses!

We have no idea when and why our ancestors lost their fur, because millennia-old mummies with hair do not exist in any sarcophagus. But we know for sure that our planet went through several periods of cold and heat. Conversely, if the climate were to blame for the loss of body hair, then fur would have to grow again at cooler temperatures. Evolution calls this adaptation.

Likewise, the opinion that primates in "our" original line started eating meat in order to be able to feed themselves better and more easily belongs in the realm of wishful thinking:. The proteins contained in meat would have given us a head start that led to intelligence. Why should it be "easier" to kill a gazelle or a fish than to grab fruit and leaves from a tree? And if meat-eating had led to intelligence faster, lions should actually have superintelligence; and crocodiles; and sharks; and all the others who have been on the hunt for some meat for millions of years.

Because a tiny reptile needed protection from the environment and from predators, protective armor was developed. This hair-raising logic demands that the genetic code, the base sequence in the DNA, has to be changed so that armor, with all the trimmings—legs, claws, and protection for the head—could be formed around the once soft animal. Why should the wish of an animal lead to the regrouping of chemical bases in the DNA? And not just any regrouping, but a targeted one. *Because* prehistoric man needed stronger teeth due to his meat consumption, they also grew for him promptly. Did the brain of this prehuman have any kind of ability to command its DNA: I need stronger teeth now? Did the DNA start to change in spermatozoa so that—Abracadabra—future generations would be equipped with model bite apparatuses for meat-eating?

To keep making the case for the evolution of man, hair-raising arguments are conjured up out of a hat. Because we (together with our ape-like ancestors) have existed for around 30 million years, we would have gained an evolutionary advantage that ultimately had to lead to intelligence. But cockroaches have existed for 500 million years, they are millions of years older than we are, and they have adapted to all adversities just as the human race did—only intelligence never evolved in them. And intelligence is much more than just adaptation. Intelligence means information exchange, culture, music, painting, tools, and technology.

Equally absurd are the assumptions that through centuries of evolution, "nature" would decide for *itself* what the particular form of life needs or what can be thrown overboard *because* it is superfluous. We, at the top of evolutionary tree, are a combination of the best of all qualities. Sorry, but we can't move as quickly as a fly. We don't have a radar system as sophisticated as bats do. Our eyes do not allow a field of vision of 342 degrees like those of a chameleon. With our fingers and hands, we cannot swing from branch to branch as elegantly and perfectly as our relatives, the chimpanzees. We cannot spit out any poisons to protect ourselves from attacks from other animals. We can't even change our color like squid. And the saddest part: we can't fly. We should—genetically!—be related to the birds, because we all originally come from the same "things."

In "nature," there should never have been a reason to give up flying, which is supposed to lead to the best adaptation. Whoever flies controls the skies. At the "top of the evolutionary tree" (or for the religious: "the crown of creation") should actually be a being able to fly and run with a panoramic view. In addition, this creature should be able to change colors in a flash, have an incredibly perfect system of sensors, and have at least a few indestructible armored areas on its body. All vital organs should be invulnerable.

But what did "evolution" do? It gave us a brain volume that we don't need; a head that can easily be destroyed and legs as a means of locomotion, which every gazelle would just find hilarious.

It's the old story: In the books on evolution, *because* acts as the key to all otherwise unsolvable problems. *Because* a form of life needed something, the adequate organ developed. Slowly, but still, it developed. The fact that certain organs cannot develop slowly should be understandable. With a wing only on one side, an animal cannot fly. So, evolution has grown two wings—one on each side of the creature—over millennia. The evolution theorists do not seem to be aware that they have to fall back on the information in the cell. *Because* the amino acids needed a protective coat, they moved into a cell. *Because* a cell needs energy, it produces chlorophyll. Each of the thousandfold *becauses* means a chemical change took place. Lifeforms like a cell do not have a brain. They would not be able to convey any wishes, which lead to so-called *becauses*, to their chemical building blocks in the cell. So there must be some reason behind amino acids changing sequences. Exactly that is what well-known scholars today call Intelligent Design.

Forty-four years ago, during a lecture at the ETH (Swiss Federal Institute of Technology) in Zurich, I met an extraordinary person: Arthur Ernest Wilder-Smith (1915–1995). He had two doctorates: one in chemistry and one in biology (Oxford University, England). Wilder-Smith authored over a hundred scientific publications. The evolutionists didn't like him because Wilder-Smith was one of those who switched sides. Originally an atheist and staunch Darwin supporter, he learned in the laboratory that the chemical building blocks in the cell did not obey any evolution or any reason. Instead, there had to be something spiritual behind them. The atheist became a Christian, and for that, his opponents used to denounce him: he had mutated into a creationist. Utter nonsense.

Wilder-Smith argued very factually. His evidence was based on research results in microbiology and genetics. The word Jesus never appeared in any of his publications or lectures. But just like Bruno Vollmert, Michael Behe, Chandra Wickramasinghe, Fred Hoyle, and countless others—some of whom I mentioned—Wilder-Smith stood by the evidence from his research. The theory of evolution was wrong. Why, when empirical evidence is available, does this finding not make its way into textbooks? Not to the public? I wonder why? The course is always set by ideology.

In Darwin's opinion, life is chemistry. Chemistry is matter. So life is a thoroughly materialistic, real matter. Translated into an ideological concept, this view is the mirror image of dialectical materialism. In addition, Wilder-Smith opines:

> *Such a materialistic theory has no place for a term like creation, cannot tolerate the supernatural and of course has no room for an indefinable power. Darwinism is the basis of their natural science and also the basis of their entire worldview, be it economic or political.*[29]

In a purely materialistic world, there is no room for a spiritual being, a so-called God or Intelligent Design, demanding accountability, a being who judges or rewards good and bad. So, using this way of thinking, it is by no means reprehensible to have millions of people killed. After all, the killers are never held accountable by anyone, and they even believe that their deeds are "good"—meaning they serve ideology. This is what happened in communism with Stalin (1878–1953), Mao Tse-tung (1893–1976) and Pol Pot (1925–1998) to the present day. There is talk of 100 million deaths for ideological reasons.[30]

Evolutionists believe—and I intentionally use the word "believe"—that their theory has been proven and that any further research is unnecessary. Occam's razor has shaved everything clean. However, the fossil evidence of an ape-like descent is more than scant and a tenuous position. In an earlier book, I pointed out that the protein structure of the apes is incompatible with ours.[31] Our genes should be 99 percent identical to those of chimpanzees—but not the proteins. For comparison: The differences in protein in two frog species are 50 times greater than those between humans and chimpanzees. Allan C. Wilson and his colleague, Mary-Claire King, both biochemists at the University of California, Berkeley, who proved these deviations in protein, said this:

> *There must have been an as yet undiscovered and, moreover, much more effective evolutionary engine than previously known.*[32]

That is true—but it's not being accepted. And in May 2004, biochemist Marie-Laure Yaspo from the Max Planck Institute for Molecular Genetics in Berlin informed the public that the differences between humans and chimpanzees are considerably greater than previously assumed:

> *Until recently, it was assumed that humans and chimpanzees differ only slightly in terms of their genetic make-up. But now a team of scientists from Germany, China, Japan, Korea and Taiwan, when directly comparing chimpanzee chromosome 22 with its human counterpart, chromosome 21, found that almost 68,000 base segments changed in the human genome; they were either added or lost . . . If these differences are extrapolated to the entire genome, apes and humans could differ in several thousand genes.*[33]

The international team of scholars also states that "the amino acid sequence of the 231 proteins discovered in humans and apes differs by 83 percent."

Actually, these results should be enough to trigger an outcry in the evolutionists' phalanx. Nothing happens, however. There's only silent yawning. Now over a hundred books have convincing arguments against Darwin's teaching. The previous "belief" turns out to be heresy, and I know very well why that is.

In my first two books, published over fifty years ago (!), I suggested an extraterrestrial influence on our evolution.[34] Earth was never a closed system. "Somebody" was always interested in getting involved. In addition to natural changes, such as those caused by radiation or chemicals, there have always been artificial mutations, targeted external interventions, throughout the history of mankind. Since our Stone Age ancestors definitely neither did genetic research nor were able to change the genetic code, only extraterrestrials remained as an explanation—exactly what today is described as Intelligent Design. In the meantime, the topic has been taken up by countless researchers and authors, but only a few have the courage to confess where their original ideas actually come from. In the US in particular, books of excellent quality have appeared, such as *The Anunnaki Connection* by archaeologist and historian Heather Lynn, or the 600-page book *Humans Are Not from Earth* by ecologist Ellis Silver.[35] He picks up a few facts that should make anyone suspicious. Here are a few examples:

All animal species on Earth need water. Without water, there is no life. All species, from beetles and lions to giraffes and hamsters, also drink unhygienic water. Who doesn't know pigs drink from every puddle? The earthly forms of life have adapted to the unclean, bacteria-infested water. But not us humans. We have to boil contaminated water. Unclean water causes diarrhea and vomiting, and can be fatal.

According to Darwin's doctrine, early humans originated in Africa. Large animals such as lions, panthers, gorillas, and crocodiles also lived there. They are all much more ideally equipped for hunting than humans.

When they fight, they react very quickly; their teeth are made up of fangs and their skin is protected by armor or thick fur. And in this dangerous world, we slowly developed into "the fittest"? In doing so, we would lose any fight against these original animals.

As beings who have adapted to this planet over millions of years, we should be able to tolerate all the food on Earth—as long as it is non-toxic. But we can't. We only eat a few types of vegetables as originally produced by nature. Animals eat the original product, but *our* stomachs have problems digesting certain types of raw vegetables. That is why we have been growing grain or corn since time immemorial in order to make it edible for us. We cook the meals and roast the raw meat—none of this is done by animals living on Earth, perfectly adapted to the environment. We cannot even eat staple foods like rice or potatoes raw. How did our ancestors even come up with the idea of growing such foods when they knew nothing about cooking? Countless people suffer from intolerance to cereals, gluten, or cow's milk. No trace of adaptation. Incidentally, it was the "gods" who taught people to grow food (I have documented this in several books). In addition, there are plant species on Earth for which we cannot prove any original form. They were just here all of a sudden. Bananas and corn are examples. I am not surprised. According to local traditions, "the gods" brought both rice and maize to Earth. When we soon set off for Mars, we will also take terrestrial plants and seeds of various foods with us.

The sun has been shining for millions of years. It would be most natural if we had adapted to it. But we didn't. Sunlight can be deadly to us. Just think of sunburn or skin cancer. Incidentally, it is not true that people with dark skin do not get cancer. They suffer from it just as often as people with light skin pigmentation.

"Evolution" gave some animals a thick coat, armor, or feathers—a natural defense layer against ultraviolet rays. Only us, the "best adapted," it abandoned. For 300,000, at best 400,000 years, early humans have been walking upright; supposedly because it gave them an advantage. Pure nonsense. An animal with four legs moves much faster than we do. Many four-legged friends can also stand up on their hind legs to get a better all-around view.

People have different skin colors. According to the theory of evolution, this different pigmentation is said to have been caused by solar influences. Originally, people were white, then black, then white again. This is documented by British evolutionary biologist Mel Greaves.[36] Greaves is an immunologist, a professor of cell biology, and the founding director the Institute of Cancer Research Center for Evolution and Cancer. He believes that the people who became hairless during the adjustment died of skin cancer. The hot sun of Africa only allowed the darker skin tones to survive. Those people who left the hot climate then promptly turned white again. The witty Armin Risi notes in his book *Evolution*:

> But the white, yellow, brown and red-skinned people are not just pale Africans, but their own types of people. In addition, the climate is the same in large parts of Europe and Asia.[37]

Communication is part of the term *language*, but language is more than just communication. Animals communicate, they exchange certain information. Something similar happens in the plant kingdom. In his work *Das geheime Leben der Bäume* (*The Secret Life of Trees*), Peter Wohlleben documents communication between plants and even between entire systems such as forests.[38] Plants warn each other about predators, but they don't speak. Before two living beings can talk to each other, they must know the meaning of the words they use. Language is more than just signs or gestures, it is more than colors or sounds; it is also more than a vibration, such as radio waves. The same goes for gestures. What does a raised, open hand mean? Friendship? Peace? Or a warning? By nodding our head, we mean agreement. But on some South Sea island, there are said to be people who mean no by nodding their head and signaling yes by shaking their head. Even giving one the "middle finger" is not understood uniformly. Language is even more complicated. In order to be able to speak to someone, both parties must be able to speak the same language. On Earth, people speak in over 6,500 different languages, not counting dialects. If I listen to a Chinese person's words but don't master their language, my ears can only pick up the melody of sounds, but I can't do anything with them. Language is precise information. Where does it come from?

After the biblical God created man and woman and they had learned to speak, they gave names to "all cattle and all birds of the sky and all animals of the field". (1. Mos. Ch. 2, verse 19 ff.)[39] A tree is not the same as a bush or branch and a lion is not the same as a hippopotamus.

At that time, as the Old Testament tells us, "all the world spoke the same language and used the same words". (1. Mos. 11, 1) In the holy Koran of the Muslims, second sura, verse 32, it says: "Thereupon he taught Adam the names of all being, showed everything to the angels and said: 'Tell me the names of these [things] if you are truthful.'"

The same can be found in the genesis myths of mankind. After some "god" had created people in his own image, he taught them to speak.[40] The Greek historian Diodorus of Sicily, who lived in the first century BC and was the author of a *40-volume historical library*, also wrote of "primitive men."[41] At first, they lived in a "half-animal state" and only banded together when they were attacked by wild animals. Their language consisted of a "mixture of sounds." Then the gods came and taught man language so that that they now "could give names to a lot of things for which there was previously no expression." In Egypt, the god Thoth taught people to speak, and in the Babylonian *Epic of Gilgamesh*—to which I will come back later—a god gave people "the reliable speech." (Plate VII, 17)[42]

A technical society is impossible without language. Even a cog is not the same as a wheel or a cogwheel. Explosives can exist in powder form, but they are neither flour nor sand nor dust, just as saliva expresses something damp, but is not water. Technologies cannot be described without very precise words, even less so mathematics. Just try to pass on a blueprint for a clock without exact expressions in a description or explain the orbit of a satellite through our solar system without wanting to calculate precise mathematics.

In a remarkable article for the magazine *Sagenhafte Zeiten* (= *Fabulous Times*), Director of Studies Peter Fiebag dealt with the mystery of the origin of language.[43] Of the millions of lifeforms on Earth, man is the only species that has the prerequisites for language. Neither beetles nor cows, not even our relatives, the apes, have the unusual larynx, without which a language cannot be implemented. This has to do with the "vocal tract above the larynx, the pharynx, mouth and nose." None of the primates have such a vocal tract. "Only humans have such an 'apparatus.'" Fiebag notes that chimpanzees don't possess this requirement; neither do any other apes and monkeys, or even other animals. This is very strange. According to Darwin's logic, the entire speaking apparatus should have developed slowly. At least our closest ancestors should have had at least a rudimentary predisposition to language. But they didn't. Fiebag writes: "Logically we would have to assume a 'development worker' here who knew exactly what the goal was."[44]

There is *one* thought running through the history of mankind that is very decisive: the virgin birth. The so-called immaculate conception describes a fertilization from *outside*—whether one calls it angels or extraterrestrials— not only in religious writings, in which every extraordinary person refers to a heavenly descent, but also in the historical texts. It begins in the more-than-5,000-year-old *Epic of Gilgamesh,* which goes back to the Sumerians. Gilgamesh—the hero of the story—is one part human and two parts divine.[45]

We continue to the scrolls found in 1947 at the Dead Sea. This also includes the Lamech Scroll. There you learn that Bat-Enosch, Lamech's wife, was artificially fertilized by one of the "Guardians of Heaven." The boy she gave birth to was Noah, the progenitor of all of us after the flood.[46] The next "heavenly birth" took place with the help of Sopranima. She was the wife of Nir, who in turn was a brother of Noah. The scroll tells how the sterile Sopranima one day gave birth to a boy who was named Melchizedech. He became the legendary priest king of the city of Salem. It is expressly pointed out that it was "an angel from heaven" who planted the seed in Sopranima's womb "without seducing her sexually."[47]

It is no different in distant Colombia. This is where the Kagaba tribe lives. A sterile woman belonging to it was fertilized by one of the "heavenly brothers." The boy she gave birth to was named Mulkueikai, and he became the progenitor of the new lineage.[48]

It goes on like this through continents and times. The tribal history of the Jaredites can be found in the *Book of Mormon*, the scriptures of the Mormons. Jared means "the one who has come down," and therefore the Jaredites derive their tribe from a divine lineage.[49]

Alexander the Great (356–323 BC) is said to have been conceived by a "lightning bolt," and the Assyrian king Ashurbanipal (687–627 BC) was a son of the goddess Ishtar. Already around 1,000 years before him, the mother of the Akkadian king Hammurabi (1728–1686 BC) was impregnated by a "sun god," and the religious founders Buddha and Zarathustra were also created by a "divine ray" in the wombs of their respective virgin mothers.

Regardless of the continent and culture—China, Central America, North and South America, Japan, the Near and Far East—and no matter what time, the Japanese great emperor named Jimmu-Tenno was just as much a descendant of "the heavenly ones" as was Gezar, the founder of the Tibetan Empire. The leading figures always referred to their heavenly descent. Now, some things may have been invented because every boss had to be something special—otherwise you wouldn't show him any respect. But Enoch—about whom I have already written many pages—even mentions the names of those "guardians of heaven" who came down to Earth and copulated with pretty human daughters.[50] And who does not know the statements in the first Book of Moses, Chapter 6?

> *But when people began to multiply on Earth . . . the sons of God saw that the daughters of men were beautiful, and they took only those as wives that they wanted.*

The statements are clear. What is confusing are the times. Any "gods" are supposed to have impregnated human women across the millennia? Is that possible? Have the extraterrestrials been present during the entire period of human creation to this day—or did they return to Earth periodically? Why? A short excursion about an animal species makes the impossible conceivable:

About 4,000 species of flies live on Earth, including the mayflies (*Ephemeroptera*; from the Greek *epheme ros* = one day and *pteron* = wing). The adult animals exist for 1 to a maximum of 4 days, others only for a few hours; the *Oligoneuriella rhenana* lives only for 40 minutes. The flies are valuable sources of data for genetic experiments. Why? If a fly only lives one hour, the species goes through 24 generations in 24 hours. This allows for very informative studies. How many generations does it take for genes to change? Is the change due to radiation or a one-sided diet? Are certain behavioral patterns passed on to the next generations? Will the fiftieth generation develop in reverse when it no longer lives in cages? In order to be able to calculate more easily, I'll take as an example a species of fly that only lives one day. After one year, that species has gone through 365 generations, after 5 human years it has gone through 1,826. A piece of cake for us— gigantic epochs for the flies. In fifteen, years, researchers in an experimental laboratory could observe over 5,400 generations of flies. Assuming that a person would live to be an average of 40 years old and that someone could observe people for 5,400 generations, this would result in an observation time of 216,000 years (5400 × 40 = 216,000).

What happened with these "gods" in the religions of Buddhism or Jainism who lived for decades and kept popping up? What about the impossible dates on the Sumerian king list? 456,000 years. What about the Mayan god Bolon Yokte, who first appeared 931,449 years ago? What about the boy-king Pakal, who held court on Earth 1,274,654 years ago?

In relation to the "gods," *we* are the mayflies. We consider our lifespans to be the measure of all things. We have no idea what age aliens reach. In addition, ETs could extend their lives indefinitely—if only by flying through space at very high speeds. Much less time passes for the crew on board the spaceship than for beings on the launch planet. This follows from Einstein's theory of relativity. These time-shifting effects have also been clearly demonstrated in experiments today.

Aliens have copulated with humans for decades. Why? They want to spread their species. They want to pass on their genes. First, they created humans "in their own likeness," then they made sure that the human species does not wither but continues to develop. With their genetic message in us, we develop into the "*Homo technicus*," who will soon be sending spaceships into the vastness of the universe. We are all no longer purely earthly beings. And since the discovery of the Lamech Scroll (Noah as the product of the "guardians of heaven"), we should have understood it, and the new knowledge should make us happy and proud. We are *more* than just "Earth animals," *more* than just the coincidental descendants from a species of ape. The "impossible" mutations that led to "*Homo technicus*" were *no* coincidences; neither were the origin of language nor the changes in the chemical bases in the DNA molecule. Everything was done in a targeted manner, and our brilliant geneticists could easily prove this—if they were allowed to publish the results of their research. Intelligent Design is the message of the alien genes in us.

With this approach, it is also understandable why we have problems with certain foods. It becomes understandable why we, in contrast to animals, can only tolerate clean water. It becomes understandable why we are more delicate than lions or crocodiles. We could not operate a computer with either the claws of a lion or those of a crocodile. It becomes understandable why we *had* to get out of the water and move onto land; the invention of electricity would not be possible under water, and without that, there would never be computers. It becomes understandable why we cannot withstand intense solar radiation—we are *not only* from this Earth. And we should also understand why people *in our time* are changing more and more. They invent the supercomputer and artificial intelligence. They learn to reject the purely materialistic world and to accept what has long been called "God" and frowned upon. Because foreign feeds into our genome are still taking place today.

I'm sorry, what?

I have documented the existence of UFOs in several books. If you don't know anything sensical about it, you should at least read Chapter 2 of my book *Botschaften aus dem Jahr 2118* (*Messages from the Year 2118*). Then you will understand the extent of the secrecy surrounding these things and also get to know leading personalities in politics, science, and the military who have made clear comments about the UFOs. On page 49 of the same book, I also took up the subject of the abduction of people by ETs.[51] I've always known how "stupid" just the thought of it is. Kidnapped by aliens? Ridiculous! I didn't want to know anything about this nonsense myself for years until I was much better informed about it as, for example, by John Mack from Harvard University (Cambridge, Massachusetts). The result? There *have been* and *there are* kidnappings of people by ETs. *There were* and *there are* genetic interventions conducted by aliens in humans. Here and today. The evidence for this is not statements by scared people whom a rash psychiatrist might classify as psychopaths; the evidence is irrefutable and empirical: implants. Such implants were discovered in the human body using computed tomography and were then removed. Whether we like it or not. (You can find the sources for this in *Messages from the Year 2118*.)

CHAPTER 4

WHERE ARE
THE FOSSILS?

The world should actually be teeming with fossils; regardless of whether Darwinists or their opponents are right. The few traces left by the opponents of the theory of evolution—footprints of shoes or human footprints in the same geological layer as the dinosaur prints—are not enough. The same is true for a slow transition from apelike creature to human. Their bones should be omnipresent. After all, people and apes have died across millions of years. Why did their bones vanish into thin air?

Not into air, but into water. The stories about one or more floods in the history of the Earth are of a global nature. (I wrote about it in detail earlier.[1]) Everyday man may have heard the story of Noah and his ark, but—cross your heart!—hardly anyone has read it in detail. The same applies to the *Epic of Gilgamesh*. None of my friends knows what is actually said there about the great flood. So let me quote three of the oldest flood stories in detail. They provide a most amazing insight into a terrible event in early human history. They also show that something still explains a lot about the missing fossils. First of all, I would like to invite readers to a story from the Bible that I will give verbatim.[2] You will learn to be amazed again. (1. Mos. Chap. 6/9 ff.)

This is the story of Noah: Noah was a pious man, blameless among his contemporaries, and he walked with God . . . Then God said to Noah: "I have decided to end all fleshly life; for the Earth is full of iniquities from men. So I will take them from the Earth. Make an ark out of fir wood: you shall make the ark with rooms, and cover it inside and out with pitch. And thus shall you make it: The length of the ark is three hundred cubits, its breadth fifty cubits, and its height thirty cubits. You shall finish it according to these measurements. You shall make a roof on top of the ark, and you shall put the door of the ark on the side. You should make a lower, a second, and a third floor inside.

But I am now allowing the flood to come upon the Earth in order to destroy all flesh under the sky that has the breath of life in it; everything that is on Earth shall perish. But with you, I will establish a covenant: you shall go into the ark, you and your sons and your wife and your daughters-in-law with you. And of all animals, of all flesh, you are to bring a pair each into the ark, in order to keep them alive with you; it shall be a male and a female. Of every kind of birds and cattle. And everything that creeps on Earth, a pair of everything should go inside to you in order to stay alive. But you take whatever food you eat and stock up on it so that you and they have nourishment. And Noah did it; just as God commanded him, exactly like that.

And the Lord said to Noah: Go into the ark, you and all your family; for I found you righteous before me among this generation. Take seven male and female of all clean animals, but one pair of the unclean animals, one male and one female, seven of the birds of the sky, male and female, so that offspring will stay alive in every part of the Earth. For after seven days, I will bring rain on the Earth, forty days and forty nights, and I will destroy all the creatures of the Earth that I have made. And Noah did exactly as the Lord commanded him. But Noah was six hundred years old when the flood came upon the Earth. And Noah went into the ark with his sons and his wife and daughters-in-law before the waters of the flood. Of the clean and unclean animals, of the birds and of everything that creeps on the Earth, a pair each, a male and a female, went into Noah's ark, as God had commanded him. And after seven days, the waters of the big flood came over the Earth. In the six hundredth year of Noah's life, on the seventeenth day of the second month, on that day all the wells of the great primordial flood erupted and the windows of heaven opened. And the rain poured onto the Earth for forty days and forty nights. On that very day, Noah went into the ark with his sons Shem, Ham and Japhet, his wife and his three daughters-in-law. They and all the various kinds of game and cattle, and everything that creeps on the Earth, and also of birds, everything that flies, and everything that has wings. They went to Noah into the ark, two each of all flesh, which had the breath of life in it. And those who went in were male and female of all flesh, as God had commanded. And the Lord locked the ark behind him.

Then the flood came over the Earth for forty days, and the waters rose and lifted the ark, and it floated high above the Earth. And the waters overflowed mightily and rose mightily over the Earth, and the ark went on the waters. And the waters became more and more mighty over the Earth, so that all the high mountains under the whole sky were covered. The waters rose fifteen cubits beyond that, so that the mountains were covered. Then all the flesh that moved on the Earth died, of birds, cattle, game and everything that swarmed on the Earth, including all human beings. Everything that breathed the air of life that was on dry land, died. So God destroyed all beings that were on the Earth: men as well as cattle, all that creeped, and birds of the sky; they were purged from the Earth; only Noah was left and everything with him in the ark. And the waters increased on the Earth for 150 days.

Then God remembered Noah and all the game and cattle that were with him in the ark. And God made a wind blow over the Earth and the waters sank; and the wells of the primordial flood and the windows of heaven closed. The rain from heaven stopped, and the waters gradually ran off from the Earth. So the waters decreased after 150 days had passed and on the seventeenth day of the seventh month, the ark settled on the mountains of Ararat. The waters continued to abate until the tenth month; on the first day of the tenth month, the peaks of the mountains became visible. Forty days later, Noah opened the window of the ark that he had made, and let the raven fly out; it flew to and fro until the waters on Earth were dried up. Noah waited seven days; then he let the dove fly out to see if the water had run off the ground. But when the dove found no place where its foot could rest, it came back again to him into the ark; for there was still water all over the Earth. Then he stretched out his hand, took the bird and put it back into the ark. Then he waited another seven days; then he let the dove fly out of the ark again. It came back to him around evening and, lo and behold, it carried a fresh olive leaf in its beak. Then Noah noticed that the waters had run off from the Earth. Then he waited another

*seven days and let the dove fly away; but it never came back
to him.*

*In the 601st year of Noah's life, on the first day of the first
month, the waters of the Earth had dried up. Noah took the
roof off the ark, and, behold, the ground was dry. On the
27th day of the second month, the Earth was quite dry. Then
God spoke to Noah, saying: Get out of the ark, you and your
wife and your daughters-in-law . . . But Noah built an altar
to the Lord; then he took of every clean animal and every
clean bird and brought burnt offerings on the altar. And
the Lord smelled the lovely fragrance and said to himself, I
will no longer curse the Earth for man's sake; the striving of
the human heart is evil from youth. And from now on I no
longer want to beat the beings that live, as I just did. As long
as the Earth exists, sowing and harvesting, frost and heat,
summer and winter, day and night shall not cease.*

*And God blessed Noah and his sons and said to them: Be
fruitful, and multiply, and replenish the Earth. The fear
and dread of you will fall on every living creature on the
Earth, every bird of the air, every creature that crawls on the
ground, and all the fish of the sea. They are delivered into
your hand. Everything that moves and lives is your food; like
the herb, the green, I give you everything . . ."[3]*

It is a fantastic story that doesn't really fit in with the
idea of an "Almighty God." He creates heaven and Earth,
plants, animals and people, "and looked at everything he
had made and, behold, it was very good." (Genesis 1:31)
But shortly afterward "the Lord repented that he had created man, and it grieved him deeply." (Gen. 6, 6) And he
decided to destroy his own creation again. Divine? At least
at the end of the story he makes a covenant with Noah and
his descendants. He promises that he will no longer curse

the Earth "for man's sake" and that sowing and harvest, frost and heat, summer and winter, day and night should "no longer cease as long as the Earth exists." However, our climate hysterics should be less happy about that. What is this "God" assuming? "Frost and heat, summer and winter" should come back year after year, and this without human intervention! What an imposition!

What additional reports do we have about other floods millennia ago?

A collection of flood sagas comes from the century before last, compiled by ethnologist Richard Andree (1835–1912).[4] Among them those of the Chaldeans, who were a Semitic people, living in southern Mesopotamia. Their writings go back to the second millennium BC. Richard Andree mentions "several Babylonian cuneiform tablets" that deal with the description of a flood, including an "eleventh book that forms the Chaldean account of the flood."[5]

This book is structured based on the signs of the zodiac. The following is the original report, from 1891, with some adjustments:

> *You know the city of Surippak, which is on the Euphrates. This city was already old when the gods urged their hearts to cause a deluge; the great gods as a whole group, their father Amu, their advisor, the belligerent Bel, their throne-bearer Adar, their leader Ennuzi. The Lord of inexplicable wisdom, the god Ea, was with them and announced their resolution to me. Man from Surippak, he said, leave your house and build a ship; they want to destroy the seed of life; therefore you shall preserve life and thereupon bring seeds of life of every kind onto the ship that you are to build. Let [indistinct number] cubits be its length and [indistinct number] be its*

width and height. Cover it with a canopy. Do not close the door of the ship behind you until I inform you. Then get on board and bring your corn, your belongings, your family, your servants and maidservants and your closest friends into the ship. I will send the cattle of the field, the game of the field to you myself.

So I built the ship and filled it with food. I divided it into sections. I checked the joints and filled them in. I poured three sars of pitch over its inside. I brought everything. I loaded onto the ship all my gold, my silver and seeds of life of every kind; all my male and female servants, the cattle of the field, the game of the field, my closest friends. When the sun god announced the appointed time, a voice said: 'In the evening the heavens will rain down. Get into the ship and close the door.' With fear I awaited the sunset. I was afraid, but I got into the ship and closed the door. I handed over the huge ship and its cargo to the Buzurkurgal, the helmsman.

Then a dark arch rose from the horizon, in the middle of which the storm god let his thunder speak. The whirlwinds are unleashed by the mighty god of the plague, the god Adar lets the canals overflow, the gods of the great, subterranean water bring up mighty floods, they make the Earth tremble, the storm god's surge rises to the sky, all light was transformed into darkness. The goddess Istar screams like a woman giving birth and cries, "so everything is turned into mud, as I prophesized to the gods. But I will not bear my sons and daughters that they fill the sea like a brood of fish." Then the gods wept with her over the spirits of the great, subterranean water.

For six days and seven nights, wind, tide, and storm prevailed. On the seventh day, however, the flood subsided, the sea withdrew into its bed, and storm and flood ceased.

But I drove through the sea, complaining loudly that the places of men had been turned into mud, the corpses drifted about like tree trunks. I had opened a hatch and when I saw the light of day, I winced, crying. I drove over the lands, now a terrible sea, when land appeared twelve measures high. The ship headed for the land of Rizir. The mountain of the land of Rizir held the ship in place. So I waited for six days. On the seventh day, I let a dove fly; since there was no resting place, it returned. Then I let a swallow fly; since there was no resting place, it returned. So I let a raven fly, and when it saw the water decrease, it did not return. I let everything out. I made a sacrifice and erected an altar on the top of the mountain.[6]

In contrast to the Old Testament, in which the story of the flood is portrayed in the third person, in the Chaldean myth the builder of the ark has a personal say. Just like in the Babylonian *Epic of Gilgamesh*. This epic was found in the hill of Kujundschik near the former city of Nineveh in what is now northern Iraq near Mosul. The *Epic of Gilgamesh* was part of the huge library of King Ashurbanipal. He ruled from 669 to 631 BC over the Assyrian Empire. But the original version of the epic is much older than Assurbanipal and goes back thousands of years. After all, *after* Assurbanipal there was no global flood, so the catastrophe reported in the *Epic of Gilgamesh* must have occurred *before* Assurbanipal's time . . .

Gilgamesh, the hero of the story, is looking for Utnapishtim, the forefather of mankind. But he lives on the other side of a great sea. During his search, Gilgamesh met some gods twice who warn him:

Gilgamesh, where are you going? You will not find the life you are looking for. When the gods created man, they determined death for man . . .[7]

After a long and adventurous journey, Gilgamesh finally finds Utnapishtim, and Utnapishtim tells him the story of the flood (eleventh plate):

> *I want to reveal to you, Gilgamesh, a hidden story, and a secret of the gods I want to tell you. Shurippak is a city on the Euphrates. It is an old city, and for a long time, the gods were gracious to it. Then the gods thought of causing a storm surge. Ea, the god of the depths of the sea, also sat in the council. He told my reed-house house the advice of the gods: "Reed-house, reed-house! Wall, O wall, hearken reed-house, wall reflect; O man of Shuruppak, son of Ubara-Tutu; tear down your house and build a boat, abandon possessions and look for life, despise worldly goods and save your soul alive. Tear down your house, I say, and build a boat. These are the measurements of the barque as you shall build her: let her beam equal her length, let her deck be roofed." I understood and I say to Ea, the God my Lord: "I will do what you command; I will obey your commandments with awe. But what should I say to the city, to the people and to the elders?" Ea opened his mouth and said to me: "You, human child, you shall say to them: Enlil, the God of Earth and countries, looks at me curiously, that's why I don't want to stay in your city anymore, I don't want to see the land of Enlil anymore. I will go down to the sea of fresh water to live with Ea, who is a gracious lord to me. But he will bless you with all kinds of riches."*

> *When the first glimmer of morning shone, I got everything ready. I went to the freshwater sea, fetched the wood and tar, sketched the ship's plan and sketched it out for myself. All my servants, strong and weak, helped with my work. The ship was completed in the month of the great Shamash. What I owned, I loaded into the ship; I brought silver and gold,*

the seeds of life of all kinds. I let women and children, my relatives and extended family board the ship. I brought in the cattle and the small animals. I let artisans of all kinds go inside.

God had told me a time to be ready: "In the evening, when the rulers of darkness pour down terrible rain, step into the ship and lock the door." The time came when Adad, the weather god, allowed terrible rain to fall. I looked at the weather; the weather was dreadful to look at. I went inside the ship and locked the door. I handed the huge boat over to the helmsman. When the morning came, pitch-black clouds rose up. All evil spirits raged, all brightness was turned into darkness. The south storm raged, the water rushed along, the water reached the mountains, the water fell upon all people. Brothers didn't recognize each other any longer. The gods themselves feared the storm surge, fled and climbed to the sky-mountain of Anu. The gods crouched down like dogs. Ishtar screams like a woman during a difficult birth, the beautiful voice of the goddess howled: "The beautiful land of yesteryear has turned to mud because I gave the assembly of gods the evil advice. How could I command such evil in the assembly of gods? How could I destroy all of my people? . . ."

For six days and six nights the rain rushed down like streams of water. On the seventh day, the storm flood subsided; it was as quiet as after a battle. The sea became calm and the storm of disaster grew still. All people were turned into mud. The soil of the Earth was a dreary monotony. I opened a hatch and the light shone in my face. I threw myself on the ground and cried, I cried, and my tears streamed down my face. I looked out over the wide open water. I shouted loudly that everyone was dead.

After twelve double hours, an island rose out of the water. The ship was drifting towards Mount Nissir. The ship ran aground and remained firmly seated on Mount Nissir. The mountain held the ship for six days and made it no longer sway; when the seventh day came, I held out a dove and let it go. The dove flew away and came back. It found no resting place, so it turned back. I held out a swallow and let it go. The swallow flew away and came back. It found no resting place, so it turned back. I held out a raven and let it go. The raven flew away, and saw the water dry up; it eats, scratches, croaks and does not turn back. So I let everyone out in all four directions and offered a lamb as a sacrifice. The gods smelled the scent, and to them it was a pleasant scent. The gods gathered like flies over the sacrifice.[8]

As the story progresses, the gods quarrel. They blame Enlil for causing the flood "without having thought about it." Enlil, for his part, cursed the others. Why had there been survivors? Who had warned the people who survived the disaster in the ark? The gods accused Enlil, "How could you so carelessly raise this flood? Let him who does sin bear his sin. Let him who commits iniquity atone for his iniquity. But see to it that not all are destroyed."

The biblical variant of the flood story is undisputedly the most recent. Someone took it from much older sources and added it to the 1st Book of Moses. From this, one can deduce that the Bible is not the only "word of God," as many people assume. In addition, the Bible speaks of Noah in the third person: ". . . and Noah did as the Lord commanded him . . . and Noah went into the ark with his sons." This is the retelling of an event. It is different in the Chaldean tradition and in the *Epic of Gilgamesh.* There, the survivor of the flood speaks in the first person: ". . . So *I* built the ship and provided it with food. *I* divided it into sections." (Chaldean) "*I* moved to the freshwater sea . . . what *I* owned, *I* loaded into the ship." (Gilgamesh).

The Bible tells you about the size of the ark. It is said to have been "three hundred cubits long" and "fifty cubits wide" and "thirty cubits tall." In the Chaldean version, the numbers for the length and breadth of the ship cannot be deciphered. Now there are several variants of the *biblical* unit of the cubit: the large or royal cubit with 52.5 centimeters (roughly 20 inches) and the normal cubit with around 44.5 centimeters (17.5 inches). (There are also other dimensions between 56 and 61 centimeters [22 and 24 inches] in circulation.) The royal cubit results in a ship length of around 157 meters (515 feet). For comparison: The *Crystal Cruises* (built in 1995) is 238 meters (780 feet) long. Noah's ark also had three floors—a truly huge ship.

The flood began "when Noah was six hundred years old, on the 17th day of the second month." Noah himself is said to have lived 950 years altogether. Neither Noah's age nor the beginning of the flood can be dated exactly. We know so little, neither when it happened nor by what standards Noah's years were counted.

The mighty ark is said to have stranded on Mount Ararat on "the seventeenth day of the seventh month," and "on the 27th day of the second month the Earth was completely dry." Since no comparative data are available, the biblical information is of no help. The Ararat mountain range, on the other hand, is well known. At 5,137 meters (close to 17,000 feet), it is the highest mountain in Turkey. Despite several expeditions, so far no evidence of the existence of Noah's Ark has emerged. In 1959, Turkish air force officer İlhan Durupınar discovered a curious formation, resembling a ship's hull, in the ground in the Ararat area, on the mountain flank of the Tendürek Dağı volcano, 27 kilometers (23 miles) from the Ararat. A ship was not found.

The same applies to expeditions in the Ararat, which were carried out in 1977 by several researchers. Neither petrified wood nor any ship planks were found. The search for an antediluvian ship via the *Epic of Gilgamesh* or the Chaldean tradition has been just as devastating. One of the cuneiform tablets of the *Epic of Gilgamesh,* which is now in the British Museum in London, gives ark (converted) dimensions of 60 × 60 × 60 meters (roughly 200³ feet). So it was a square block with a volume of 216,000 cubic meters (over 7,500,000 cubic feet). This means that this transport vehicle was five times larger than the biblical ark.[9] Despite the exciting searches of various research teams who set out on the trail of this incredible ship, not a single piece of real evidence has turned up to this day.[10]

The *Epic of Gilgamesh* even mentions the father of Utnapishtim. His name was Ubara-Tutu. That doesn't match the biblical version. There Noah would be Utnapishtim, but Noah's father was called Lamech and not Ubara-Tutu. The information about the landing place is just as contradictory. In the Bible, the ark is stranded "on Mount Ararat," in the Chaldean variant "on Mount Nissir" and in Gilgamesh on the "mountain of the land of Rizir." It is entirely possible that all authors meant the same region, but mountains have different names for different peoples. A well-known mountain in the Swiss canton of Valais is called the Matterhorn— in Italy, however, the same mountain is called Cervino. For German speakers there is a Lake Geneva, for French speakers it is called Lac Léman. The sumerologist Hermann Burgard writes:

> *In the original text of the Bible, in the first book of Moses, chapter 8.4, in German version it says this, translated: ". . . put*

the ark on the Ararat mountains." That is an unauthorized
narrowing by the translators. It should have been worded like
this: ". . . touched down on the mountains of rrt." The vowels
are missing in the original. It could mean Ararat or the wider
area of Urartu.[11]

In Noah's case, a pair of each animal species is taken
on board—in the other traditions, however, "seeds of life
of all kinds" (Chaldean) and "craftsmen of all trades."
Obviously it was about a worldwide flood, but afterward
Earth was to turn green again and craftsmen—from car-
penter, smith, bricklayer, and farmer to scribe—had to be
protected. It was not about a new start from scratch; the
"gods," those who caused the flood, had planned very well
in advance.

Why should aliens even kill their own breed ("the
gods created man in their own image")? Because the
first experiment went wrong. Not only did the pre-flood
people, but also some of the ETs disobey the rules, but
a brood of undesirable lifeforms developed. In biblical
terms: the seduction of Adam and Eve by "the serpent."
The undesirable lifeforms had begun to spread across
Earth and could no longer be tracked down and killed
individually. (In my book *Falsch Informiert!* [*Wrongly
Informed!*] I dealt with the process in detail.[12]) The decisive point is this: The goal of the great clean-up was to
allow a new gene pool to grow. And indeed: the few survivors after the flood carried a *modified* DNA. We humans
of the twenty-first century are the descendants of those
forefathers. And since this flood, new genetic information
is repeatedly fed into the human system. The keyword
is virgin birth. *Because of this*, geneticists can no longer
explain the mutations in the chemical building blocks in

our DNA in a natural way; because they did *not* develop by themselves à la Darwin. This is what we call Intelligent Design in our time.

But, as I hear in every discussion, the flood, if it occurred at all, was a locally restricted event of completely natural forces. In fact, there were geographically restricted floods at all times, easily detectable in the respective strata of Earth. But that primordial flood of human tradition was neither about a local natural event nor about the punishment by a spiritual being called "God."

Why not? An "almighty" god could have conjured up a ship to save his protégés with the snap of his fingers. But he was incapable of that. In every description of the flood, a "god" orders a selected group of people to build a ship. Shipbuilding is technology. Trees have to be felled and boards cut. Plans have to be drawn up; strings, ropes, and pitch have to be brought in. It all takes time. Whoever warned his chosen people knew the time of the flood. And he knew that this tide would not come in two weeks, but only in a few months. Otherwise there would not have been enough time to build the ship.

So this could not have been a natural event. The flood was planned.

The objection that it was a local event is taken ad absurdum by the traditions (for the umpteenth time—I wrote about it earlier and therefore only cite a section from my last book *Confessions of an Egyptologist*):[13]

> *Whether it was the Greek geographer Strabo (around 63 BC to 23 AD) or Pliny the Elder (23–79 AD), whether it was Hesiod (around 700 BC) or Herodotus (486–430 BC), whether it was Hekataios (approx. 560–480 BC) or the Babylonian Berossus (late 4th century BC), whether it*

was Diodorus of Sicily (1st century BC) or the unknown Maya writers in distant Central America, it doesn't matter who I call as a witness. Because the fact always remains the same; all wrote about floods that occurred 10,000 years ago and more. The Greek philosopher and scholar Plato reported about it just like an unknown writer did in the distant land of Sumer millennia before him. On a cuneiform tablet, translated by the famous sumerologist James Pritchard (1909–1997), which is now in the Baghdad Museum, one can read:

"I want to trigger a flood. It will wipe out the seeds of the people . . . The flood will wipe out all centers . . . the flood will destroy the countries."[14]

The claim that one of these early historians copied the other is a joke. The unknown scribblers of the Maya in Central America could not have known anything about the Greek philosopher Plato. He lived from 428 to 348 BC on a different continent than the Maya. There was no contact. But both described the flood. And the indigenous tribe of the Kabeza in the highlands of Colombia knew nothing about Sumerian cuneiform tablets and a Mr. Gilgamesh. But both reported a worldwide flood. Despite the millions of lost books in the past, Earth's history has been known from 500 BC up until now. At least for those who know Greek history. We definitely know that no flood has destroyed the entire globe in the past 2,500 years. So, the historians of prehistoric times describe a flood that must have happened more than 2,500 years ago. At the time of the pharaohs in Egypt, there was no flood either. The Egyptian temples prove it. So it must have happened *still* earlier. And you inevitably end up in the haze of millennia—otherwise neither a

Chaldean nor the author of the *Epic of Gilgamesh* would have been able to write about it.

A worldwide flood that even covered the mountains not only destroyed graves and fossils, but also all ancient cultural centers and their temples. That is why the grandiose structures that we marvel in areas of the Maya, in India and Egypt were all built *after* the flood.

One of the few exceptions to this are the great pyramids in Egypt. They were built *before* the flood. The builders were informed about the coming flood—that's why the pyramids were built in the first place. They *were meant* to survive the waters. (All about this my book *Neue Erkenntnisse.*[15])

I have been grappling with the riddles of early human history since my youth. For many of these puzzles, there are now reasonable explanations. What annoys me again and again in all discussions is this: right through the literature of all cultures, people testified that they were created by the gods. Only the materialistic, conceited, and high-handed people of today reject this indignantly. There must be no "gods" and certainly nothing "spiritual." We came into existence by ourselves! We did everything ourselves! We are the greatest! Images sent to Earth by NASA satellites from the surface of Mars clearly show that there are traces of former artificial structures there (See Figure 4.1 and Image 14 in the color insert). Since we humans have never been to Mars, the ruins must stem from someone else. Is that understood?

Figure 4.1: Mars anomaly

NOTES

CHAPTER 1

1. Sabrina Schröder, "10 Parasiten, die Tiere zu Zombies machen [10 parasites, the animals that make zombies]," *Spektrum der Wissenschaft* [Spectrum of Science], September 2, 2019.
2. Charles Darwin, *Die Entstehung der Arten* (*On the Origin of Species*) (Stuttgart, Germany: 1974).
3. William Shearer, *The Atlantic Salmon* (New York: 1992).
4. Carsten Jasner, "Meilensammler," P. M., January, 1, 2020, 38–45.
5. Kenneth C. Catania, "The Astonishing Behavior of Electric Eels," Frontiers in Integrative Neuroscience, July 16, 2016, *https://doi.org/10.3389/fnint.2019.00023*.
6. Alexander von Humboldt, *Reise in die Aequinoctial Gegenden des neuen Continents* (Stuttgart: 1859).
7. Nick Stringer, *Tortuga*, 20th Century Studios.
8. Nico Michiels, "Zwitter spielen gerne Männchen," *Innovations Report*, June 25, 2007, *https://www.innovations-report.de*.
9. J. Emmett Duffy and Kenneth S. Macdonald "Kin Structure, Ecology and the Evolution of Social Organization in Shrimp," *Proceedings of the Royal Society B: Biological Sciences* 277, no.1681 (February 22, 2010): 575–584, *https://doi.org/10.1098/rspb.2009.1483*.
10. Erich von Däniken, *Botschaften aus dem Jahr 2118* (Rottenburg, Germany: 2016).

11. Petr Necas, *Chamäleons. Bunte Juwelen der Natur* (Frankfurt, Germany: 2004); Wolfgang Schmidt, Klaus Tamm, and Erich Wallikewitz, *Chamäleons, Drachen unserer Zeit* (Münster, Germany: 2010).

12. Gaius Plinius Secundus, *Die Naturgeschichte* (Leipzig, Germany: 1882).

13. Ant, *https://en.wikipedia.org*.

14. Ameisen, *https://de.wikipedia.org*.

15. Christina Grätz, *Im Königreich der Ameisen*, ARTE-TV, April 18, 2020.

16. Christina Grätz and Manuela Kupfer, *Die fabelhafte Welt der Ameisen* (Gütersloh, Germany: 2019).

17. William H. Gotwald, *Army Ants: The Biology of Social Predation* (New York: 1995).

18. Walther Kirchner, *Die Ameisen. Biologie und Verhalten* (München, Germany: 2007).

19. Ameisen, *https://de.wikipedia.org*.

20. Termiten, *https://de.wikipedia.org*.

21. Ulf Hohmann und Ingo Bartussek, *Der Waschbär* (Einbeck, Germany: 2018).

22. Waschbär, *https://de.wikipedia.org*.

23. Schuppentiere, *https://de.wikipedia.org*.

24. Schuppentiere, *https://de.wikipedia.org*.

25. Hundertfüsser, *https://de.wikipedia.org*.

26. *https://www.ripleybelieves.com*.

27. Max Daunderer, *Klinische Toxikologie in der Zahnheilkunde* (1998), and *Lexikon der Pflanzen und Tiergifte* (1995).

28. Matthias Bergbauer, *Giftige und gefährliche Meerestiere*, (Rüschlikon, Switzerland: 1997).

29. Ant, *https://www.meerwasserlexikon.de*.

30. Dieter Mahsberg, Rüdiger Lippe, and Stephan Kallas, *Skorpione* (Münster: 1999).

31. Gary Polis, *The Biology of Scorpions* (Los Angeles: 1990).

32. Frank Nischk, *Die fabelhafte Welt der fiesen Tiere* (Regensburg, Germany: 2020).

33. Nischk, *Die fabelhafte Welt*.

34. *https://www.travelbook.de*.

35. Rudolf Diesel and Christoph D. Schubart, "Die außergewöhn-liche Evolutionsgeschichte jamaikanischer Felsenkrabben," *Biologie in unserer Zeit* 30, no. 3 (2000): 136–147.

36. Helmut Debelius, *Krebs-Führer Weltweit* (Hamburg, Germany: 2000).

37. *https://de.wikipedia.org/wiki/Fangschreckenkrebse #Lebensweise.*

38. Vogelzug, *https://de.wikipedia.org.*

39. Vogelzug, *https://de.wikipedia.org.*

40. Peter Berthold, *Vogelzug. Eine aktuelle Gesamtübersicht* (Darmstadt, Germany: 2008).

41. Volker Mrasek, "Dieser Vogel fliegt dreimal zum Mond," *NZZ Magazin,* May 1, 2020.

42. Peter Matthiessen, *Die Könige der Lüfte. Reisen mit Kranichen* (München: 2007); Günter Nowald and Hermann Dirks, *Kranichbegegnungen Kranichwelten* (Dresden, Germany: 2006).

43. Peter Godfrey-Smith, *Other Minds: The Octopus and the Evolution of Intelligent Life,* (Zürich, Switzerland: 2018).

44. Benjamin P. Burford and Bruce H. Robison, "Bioluminescent Backlighting Illuminates the Complex Visual Signals of a Social Squid in the Deep Sea," *PNAS* 117, no. 15 (March 23, 2020): 8524–8531, *https://doi.org/10.1073/pnas .1920875117.*

45. Johann Grolle, "Was hinter der Intelligenz der Tintenfische steckt," *Der Spiegel,* December 20, 2017, https://*www .spiegel.de.*

46. Andreas Hirstein and Theres Luthi, "Crispr-Cas9: So arbeitet die Gen-Schere," *NZZ magazine,* February 7, 2016, *https://magazin.nzz.ch.*

47. Johann, Grolle, "Lenkwaffe im Zellkern," *Der Spiegel* 18, April 24, 2015, *https://www.spiegel.de.*

48. Isabel C. Vallecillo-Viejo et al., "Spatially Regulated Editing of Genetic Information within a Neuron," *Nucleic Acids Research,* 48, no. 8 (May 7, 2020): 3999–4012, *https://doi .org/10.1093/nar/gkaa172.*

49. Braem, Guido J., *Fleischfressende Pflanzen* (München: 2002); Adrian Slack, *Karnivoren: Biologie und Kultur der insektenfangenden Pflanzen* (Ulm, Germany: 1985).

50. Venusfliegenfalle, *https://de.wikipedia.org.*

51. Mitsch, Jacques, *Der Blob—Schleimiger Superorganis mus,* TV-Sender ARTE, March 18, 2020; *https://www.welt.de.*

52. Wolfgang Richter, "Alter Schleimer," *Die Zeit,* December 29, 2006.

53. Mitsch, *Der Blob.*

54. Richter, "Alter Schleimer."

CHAPTER 2

1. Rudolf Hoffmann, *Das Neue Organon* (Leipzig, Germany: 1962).

2. Francis Bacon, *Meditationes Sacrae and Human Philosophy,* 1597.

3. Otfried Höffe, *Immanuel Kant* (München, Germany: 2007).

4. Manfred Geier, *Kants Welt* (Reinbek: 2003).

5. Otfried Höffe, *Kants Kritik der reinen Vernunft* (München: 2000).

6. Sabine Appel, *Arthur Schopenhauer. Leben und Philosophie* (Düsseldorf, Germany: 2007).

7. Arthur Schopenhauer, *Die Welt als Wille und Vorstellung* (Köln, Germany: 1997).

8. Ernst Haeckel, *Natürliche Schöpfungsgeschichte,* available from *http://www.zum.de.*

9. Jean E. Charon, *Der Geist der Materie* (Hamburg, Germany: 1979).

10. Erika Krausse, *Ernst Haeckel* (Leipzig: 1984).

11. Bernhard Kleeberg, *Theophysis. Ernst Haeckels Philosophie des Naturganzen* (Köln: 2005).

12. Haeckel, *Naturliche Schopfungsgeschichte.*

13. Friedrich Nietzsche, *Gesammelte Werke; Zur Genealogie der Moral; Der Fall Wagner* (Zürich, Switzerland: 1985).

14. Nietzsche, *Gesammelte Werke.*

15. Nietzsche, *Gesammelte Werke.*

16. Arthur Keith, *An Autobiography* (London, 1950).

17. *https://www.soulsaver.de.*

18. *https://www.soulsaver.de.*

19. Malcolm Muggeridge, *https://en.wikiquote.org.*

20. Michelle, *Wildlife of Greater Brisbane und Wildlife of Tropical North Queensland,* (Brisbane, Australia: 2003).

21. Reinhard Junker, and Siegfried Scherer, *Evolution: Ein kritisches Lehrbuch* (Gießen, Germany: 1998).

22. *https://www.spiegel.de.*

23. Martin Brookes, *Drosophila. Die Erfolgsgeschichte der Fruchtflie,* (Hamburg: 2002).

24. Richard Dawkins, *Der Gotteswahn (The God Delusion)* (Zürich: 2008).

25. Jan P. Beckmann, *Wilhelm von Ockham* (München: 1995); Volker Leppin, *Wilhelm von Ockham: Gelehrter, Streiter, Bettelmönch* (Darmstadt, Germany: 2003).

26. Erich von Däniken, *Botschaften aus dem Jahr 2118* (Rottenburg, Germany: 2016), Chapter 3.

27. Jacques Monod, *Zufall und Notwendigkeit* (München: 1975).

28. Manfred Eigen, *Das Spiel—Naturgesetze steuern den Zufall* (München: 1975).

29. Michael, J. Behe, *Darwin's Black Box* (Gräfelfing, Germany: 2007).

30. Chandra Wickramasinghe, "Die Entdeckung außerirdischen Lebens—Ein Wendepunkt in der Geschichte der Menschheit," Speech given at the International Erich-von-Däniken-Kongress, 2015.

31. Fred Hoyle, *Das intelligente Universum* (Frankfurt: 1984).

32. Fred Hoyle and Chandra Wickramasinghe, *Evolution aus dem All* (Frankfurt: 1981).

33. Arthur D. Horn, *Götter gaben uns die Gene* (Berlin: 1997).

34. Bruno Vollmert, *Das Molekül und das Leben* (Hamburg: 1985).

35. Thomas Nagel, *Mind and Cosmos: Why the Materialist Neo-Darwinian Conception of Nature Is Almost Certainly False* (Oxford, England, 2012).

36. Behe, *Darwin's Black Box.*

37. Behe, *Darwin's Black Box.*

38. Michael Behe, *https://rationalwiki.org.*

39. Dawkins, *Der Gotteswahn (The God Delusion).*

40. Behe, *Darwin's Black Box.*

41. Behe, *Darwin's Black Box.*

42. Reinhard Junker, über: *https://www.genesisnet.info*.

43. Erich von Däniken, *Neue Erkenntnisse* (Rottenburg: 2018), 173–206.

44. John E. Mack, *Abduction: Human Encounters with Alien* (New York: 1994); Illobrand von Ludwiger, *Ergebnisse aus 40 Jahren UFOForschung* (Rottenburg: 2015); Johannes Fiebag, *Kontakt. UFOEntführungen in Deutschland, Österreich und der Schweiz* (München: 1994).

45. Däniken, Erich von, *Botschaften aus dem Jahr 2118* (Rottenburg: 2016).

46. Behe, *Darwin's Black Box.*

47. Dawkins, *Der Gotteswahn* (*The God Delusion*).

48. William Paley, *Natural Theology* (London, 1802).

49. Behe, *Darwin's Black Box.*

CHAPTER 3

1. Luis Navia, *Unsere Wiege steht im Kosmos* (Düsseldorf, Germany: 1976).

2. Oliver Mühlemann, "Wie LUCA, die Urzelle des Lebens, entstand," *UniPress* 157 (University of Bern, Switzerland: 2013): (13–15).

3. Michael A. Cremo and Richard L. Thompson, *Verbotene Archäologie* (Rottenburg, Germany: 2006).

4. Cremo and Thompson, *Verbotene Archäologie.*

5. Wilbur G. Burroughs, "Human-Like Footprints, 250 Million Years Old," *The Berea Alumnus* (Berea College, KY: November 1938).

6. Burroughs, "Human-Like Footprints."

7. Albert G. Ingalls, "The Carboniferous Mystery," *Scientific American* 162, no. 1 (January 1940).

8. Erich von Däniken, *Botschaften aus dem Jahr 2118* (Rottenburg, 2016).

9. Cremo and Thompson, *Verbotene Archäologie.*

10. Helmuth von Glasenapp, *Der Jainismus: Eine indische Erlösungsreligion* (Hildesheim, Germany: 1984; originally published in 1925); Erich Frauwallner, *Geschichte der indischen Philosophie* (Salzburg, Austria: 1953).

11. David Stuart, and George Stuart, *Palenque: Eternal City of the Maya* (London: 2008).

12. Hans-Joachim Zillmer, *Die Evolutions-Lüge* (München, Germany: 2005).

13. Erich von Däniken, *Beweise* (Düsseldorf, Germany: 1977), 324 ff.

14. Brad Steiger, *Mysteries of Time and Space* (New York: 1974).

15. "Saurier und Primaten lebten Seite an Seite," *Die Welt* June 14, 2004, 35 (with a reference to an article in *Nature*).

16. Steiger, *Time and Space*, 292.

17. Daniel Gerber, "Kalt Wind für Darwin-Theorie: Über 1,000 Wissenschaftler unterzeichnen Kritik-Petition, Livenet.ch, February 18, 2019, *https://www.livenet.ch*.

18. Pedro Lima, "Sternstunde der Steinzeit," *Focus Magazine*, no 50/2000 (December 18, 2015).

19. Erich von Däniken, *Die Götter waren Astronauten!* (Rottenburg: 2015).

20. Lima, "Sternstunde der Steinzeit."

21. Erich von Däniken, *Odyssey of the Gods: The History of Extraterrestrial Contact in Ancient Greece* (Newburyport, MA: 2011).

22. Thorsten Morawietz, "Versunkene Ruinen vor Malta," *Sagenhafte Zeiten*, no. 6 (2019).

23. Srila Vyasadevas, *SrimadBhagavatam*, trans. A. C. Bhaktivedanta Swami Prabhupada (Vienna, Austria: 1987).

24. S. R., Rao, *The Lost City of Dvaraka* (New Delhi, India: 1999).

25. Rao, *Lost City of Dvaraka*.

26. Zillmer, *Die EvolutionsLüge*, 302.

27. "In vollem Gang: Das sechste Artensterben," *Welt am Sonntag*, no. 24, (June 14, 2020) (under a reference from an article in the *Proceedings of the National Academy of Science [PNAS]*).

28. "In vollem Gang," *Welt am Sonntag*.

29. Arthur. E. Wilder-Smith, A. E., *Die Erschaffung des Lebens*, (Stuttgart: 1972).

30. Stéphane Courtois, *Das Schwarzbuch des Kommunismus* (München: 1998).

31. Däniken, *Botschaften*.

32. "Biologie: Durch Gen-Rutsch zum nackten Affen," *Der Spiegel*, no. 18 (April 27, 1975).

33. International Chimpanzee Chromosome 22 Conortium, "DNA Sequence and Comparative Analysis of Chimpanzee Chromosome 22,"*Nature* 429(2004): 382–388.

34. Erich von Däniken, *Erinnerungen an die Zukunft* (Düsseldorf, 1968), 96; Erich von Däniken, *Zurück zu den Sternen* (Düsseldorf: 1969), 33 ff.

35. Heather Lynn, *The Anunnaki Connection. Sumerian Gods, Alien DNA, and the Fate of Humanity* (Newburyport, MA: 2020); Ellis Silver, *Humans Are Not from Earth* (Cullompton, UK: 2017).

36. Mel Greaves, *Krebs—der blinde Passagier der Evolution* (Hamburg, Germany: 2002).

37. Armin Risi, *Evolution: Stammt der Mensch von den Tieren ab?* (Zürich: 2014).

38. Peter Wohlleben, *Das geheime Leben der Bäume* (München: 2016).

39. *Die Heilige Schrift des Alten und des Neuen Testaments Württembergische Bibelanstalt* (Stuttgart, 1972).

40. Däniken, *Botschaften*, 127 on.

41. Adolf Wahrmund, *Diodor's von Sicilien Geschichts Bibliothek*, 1. Buch (Stuttgart, 1866).

42. Wahrmund, *Diodor's von Sicilien Geschichts Bibliothek.*

43. Peter Fiebag, "Phänomen Sprache," *Sagenhafte Zeiten*, no. 5 (2009).

44. Fiebag, "Phänomen Sprache."

45. Georg Burckhardt, *Gilgamesch. Eine Erzählung aus dem alten Orient* (Wiesbaden, Germany: 1967).

46. Millar, Burrows, *Mehr Klarheit über die Schriftrollen* (München: 1958).

47. Gottfried Wuttke, *Melchisedech, der Priesterkönig von Salem,* (Gießen, Germany: 1929).

48. Gerardo Reichel-Dolmatoff, "Die Kogi in Kolumbien," *Bild der Völker*, Band 5, Wiesbaden o. J.

49. *Das Buch Mormon*, Buch Ether.

50. Erich von Däniken, *Falsch informiert!* (Rottenburg: 2007), 53 ff.

51. Däniken, *Botschaften*, 49.

CHAPTER 4

1. Erich von Däniken, *Im Namen von Zeus* (München, Germany: 1999).

2. *Die Heilige Schrift des Alten und Neuen Testaments* (Stuttgart, Germany: 1972).

3. *Die Heilige Schrift*, 1 Mos. 6: 13–22, 7: 1–24, 8: 1–22, 9: 1–3.

4. Richard Andree, *Die Flutsagen. Ethnographisch betrachtet* (Braunschweig, Germany: 1891).

5. Andree, *Die Flutsagen.*

6. *Die Heilige Schrift.*

7. *Epic of Gilgamesh.*

8. *Epic of Gilgamesh.*

9. Arche_Noah, *https://de.wikipedia.org.*

10. Charles Berlitz, *Die Suche nach der Arche Noah* (Hamburg, Germany: 1987).

11. Hermann Burgard, *Flutheld Ziusudra,* (Groß-Gerau, Germany, 2020).

12. Erich von Däniken, *Falsch informiert!,* (Rottenburg, Germany: 2007), 53 ff.

13. Erich von Däniken, *Die Bekenntnisse des Ägyptologen Adel H.,* (Rottenburg: 2019), 33.

14. James B. Pritchard, *Ancient Near Eastern Texts* (Princeton, NJ: 1955).

15. Erich von Däniken, *Neue Erkenntnisse* (Rottenburg: 2018), 173–206.

IMAGE SOURCES

FIGURES IN THE TEXT

Figures 1.1–1.24: Wikimedia Commons
Figure 1.25: David Müller, Bern
Figures 1.26–2.8: Wikimedia Commons
Figure 3.1: W. J. Meister, Antelope Springs, USA
Figures 3.2 and 3.3: Ramon Zürcher, Unterseen
Figure 4.1: Taken by Mast Camera (Mastcam) onboard NASA's
   Mars rover Curiosity on Sol 1438 (2016-08-22 11:37:56
   UTC) Image Credit: NASA/JPL-Caltech/MSSS

COLOR INSERT IMAGES

Images 1–13: Wikimedia Commons
Image 14: https://ida.wr.usgs.gov/html/e10004/e1000462
   .html

ABOUT THE AUTHOR

ERICH VON DÄNIKEN is arguably the most widely read and most copied nonfiction author in the world. He published his first (and best-known) book, *Chariots of the Gods*, in 1968. The worldwide bestseller was followed by more than three dozen books, including the recent bestsellers *Confessions of an Egyptologist, War of the Gods, Eyewitness to the Gods, The Gods Never Left Us, Twilight of the Gods, History Is Wrong, Evidence of the Gods, Remnants of the Gods,* and *Odyssey of the Gods.* His works have been translated into twenty-eight languages and have sold more than sixty-five million copies. Several have also been made into films. von Däniken's ideas have been the inspiration for a wide range of television series, including the History Channel's hit *Ancient Aliens.* His research organization, the AAS-RA/legendarytimes.com (Archaeology, Astronauts, and SETI Research Association), comprises laymen and academics from all walks of life. Internationally, there are about 10,000 members. Erich lives in Switzerland but is an ever-present figure on the international lecture circuit, traveling more than 100,000 miles a year.

To follow the latest visit *www.daniken.com/en/* or Erich von Däniken's Official Fan Page on Facebook.